It's Not Magic, It's Biology:

a guided tour through

your molecular world

by

Allan Albig

with illustrations by

Roxanna Albig

Dedication:

Of all the people that might be recognized here, it is my sister, Jenny Leaser, to whom this book is dedicated. As I wrote the book, I kept asking myself if my non-scientist sister would understand what I was trying to say. If the answer was "no", I tried again. Hopefully, these chapters will help you think about molecular biology and how it applies to your world.

I also want to thank Warren J. Hagestuen, who was my first biology teacher in 7^{thish} grade, for starting me down this path in the first place.

Contents

Preface

As a professional molecular biologist, one of the most difficult questions I get asked when I meet a new person is "what do you do for a living?" Here is how the conversation usually goes:

> "Good to meet you Allan, what do you do for a living?"

> "I am a professor at Boise State University. I teach classes in cell biology and molecular genetics and run a research lab."

> "That sounds fascinating, what do you work on in your lab?"

> "I study a cell communication pathway called Notch. It's basically a way that two cells can use proteins on their cell surfaces to make contact and communicate with each other about what they are doing."

> Head nodding politely. "Oh, that sounds interesting. Done any fishing lately?"

After years of having this conversation, and of immediately losing my audience, I've realized that many people don't care to talk about how cells, much less how molecules, work. It could be that they don't find it interesting, but more often I think it's that they have no reference points from which they can construct an understanding of what happens in my lab. Or maybe they think it's so confusing there is no point in talking about it. This isn't surprising since molecular biology is rarely discussed, even in popular science. Myriad great science articles about fascinating topics are published every day. Many of them could easily introduce and discuss some simple molecular mechanisms, yet the vast majority of popular science articles only vaguely hint at the fascinating molecular biology that is just under the surface.

I have also noted that there are very few resources designed to teach non-scientists about how molecules work in living systems. Almost every molecular biology book I have ever seen, with the exception of *The Manga Guide to Molecular Biology* and maybe some of Larry Gonick's *Cartoon Guide* books, have been written for upper level college students in advanced biology classes. This is unfortunate. Molecular interactions are the foundation of our everyday life; individual molecules shape and govern how we perceive every single experience we have, even though we cannot even see them. And yet, most people don't understand the simplest principles of molecular behaviors.

I think this lack of understanding is unnecessary. While molecular interactions can become very complex, I assure you that you don't have to be a professional molecular biologist to appreciate the simple elegance of what molecules do and how they do it. I think that a very fundamental understanding of molecular biology can enrich anyone's life, and should be accessible to everyone!

So I decided to try to fix the problem in the way that most professors do, which is to say, I wrote a book. **The goal of this book is to help people understand how biology works at a molecular level and to help them appreciate the molecular world**.

It may sound funny, but I am indebted to electric eels since these creatures gave me the idea that sparked this book. As I taught students about electric eels, I thought to myself, "how could anybody **not** be interested in knowing and understanding how an electric eel produces several hundred volts of electrical energy and uses it to communicate and capture food?!" If you don't understand it, it might seem like magic. But here is the really cool thing, if you **do** understand how eels generate this voltage, it's even **more** amazing than magic. These "magical" abilities that living things have are built on very simple, yet highly effective, systems that nearly everyone can understand. My challenge, of course, is to help you really appreciate how amazing that eel is, even if you don't have a background in cell and molecular biology.

Appreciation of an idea or a phenomenon is a real gift. My wife is a professional musician and has tried to get me to **really** appreciate what she calls Bach's *cosmic music.* I don't have a music degree. My playlists are filled with heavy metal music. I acknowledge that Bach sounds nice, but it's difficult to realize the true beauty of Bach's work since I don't have the knowledge that would allow me to fully appreciate his achievement. With a little bit of handholding by a professional who can help me see the music from a trained perspective; I have gained some level of appreciation for what's happening with the music without needing to go back to college again. This appreciation opens up and deepens my understanding of the world. That's what I wanted to do with this book; to give non-scientists an appreciation for what's going on in the molecular world and how that world works. So, let me hold your hand and try to show you that the molecular world is not magic, it's just biology.

And also, for the record, the fishing has been great lately.

Introduction

Who cares about molecular biology?

In my opinion, the first and best reason to understand at least a little molecular biology is that it will change the way you look at the world. It will give you a better appreciation for all life. Even murder hornets. The second reason is that, whether you know it or not, you depend on molecular biology to get through every single day and having a little understanding will help you to navigate your world.

Don't believe me that molecular biology impacts your everyday life? Here is an obvious example from 2020: the COVID-19 pandemic. How can this little invisible virus do so much harm to so many people and to the global economy? The simple answer is rooted in molecular biology. Viruses are super elegant, compact packages of minimalist molecular biology that have only one mission in "life": to make more viruses just like them. But the virus needs a host, and in order to get inside that host, the virus needs something in the host that it can stick to. The SARS-CoV-2 virus uses a sticky protein on its surface, aptly named the *spike protein,* that tightly sticks to the *Ace-2 protein*, which is produced in your body and presented on the outside of some of your cells. The interaction is akin to molecular Velcro. When the spike protein binds to Ace-2, the binding interaction fools your own cell into pulling the virus inside of itself where the virus can complete its mission. **This simple interaction between two proteins based on matching shapes and chemistries is one of the most fundamental ideas in molecular biology and is arguably the root of all the devastation of the viral pandemic.**

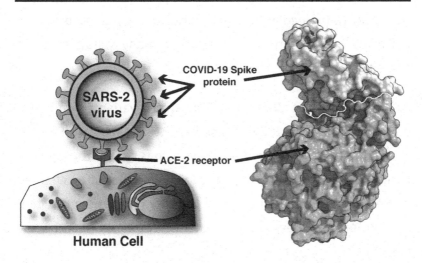

Figure 1: Sars-2 spike protein binding to its receptor, ACE2. The SARS-CoV-2 spike protein binds to the ACE-2 receptor on the surface of cells in your lungs and blood vessels. The figure on the right shows the actual interaction between the spike protein and its receptor. this interaction is only possible because of the perfectly matched shapes and chemistries of these two proteins. This is a fundamental concept in molecular biology. Image on the right is taken from open access paper [1].

What about the new vaccines that have been approved to prevent the spread of COVID-19? The first vaccines are called mRNA vaccines and they work through molecular mechanisms that are shared by all life on Earth. These mRNA vaccines are different than traditional vaccines. When the mRNA vaccine is injected into your arm, your cells read the mRNA code, produce the same spike protein that the virus uses to attach to cells, and presents that spike protein to your immune system. Your immune system recognizes that the spike protein is a "foreign protein" that is not supposed to be in the body and starts making antibodies, which will attack any other spike proteins they encounter. We only know how this works because of all the effort of molecular biologists, and we only learned how to make and use these mRNA vaccines after many years of studying these basic biological mechanisms.

Finally, what about all the new SARS-CoV-2 variants that began popping up in early 2021? These variants are new SARS-CoV-2 mutants that have evolved as the virus has spread through the human population. Some of these mutants have subtle changes in their spike proteins that could possibly make them resistant to our new vaccines, although this does not seem to be the case, yet.

So why do **you** need to understand molecular biology? Because it makes you prepared to understand the world around you. You will be less dependent on what other people tell you, and more empowered to make up your own mind about what is true or what is just somebody's online opinion. We are bombarded by information every day and having some basic skills to sort the truth from the "fake news" has never been more important than it is right now. For instance, a question that you will undoubtedly face is whether or not you are going to take the COVID-19 vaccine. Whether you do or don't take the vaccine is your choice, but if you understand a little about how the molecular biology of the vaccine works, you will be in a better position (with your doctor's advice of course) to confidently make the choice that is right for you.

I could go on and on describing how molecular biology impacts your every day. I don't think you want that though. You already picked up this book so I have to assume you are ready and eager to learn, and that you don't need me to try to convince you that you should! So, let's get to it!

Molecular Biology 101

In the first day of your first high school biology class, your teacher probably taught you that "bio" means life and that "ology" is the study of something. Putting this together then, biology is the study of life. Okay then, what is molecular biology? It's the study of the molecules that all life depends on. The study of the molecules of life. No doubt you have heard of these molecules already even if you didn't appreciate them as molecules of life. They are proteins, carbohydrates, fats or lipids, and the nucleic acids, DNA and RNA. Molecular biology seeks to understand all the complexities of how these molecules work.

Let's talk specifically about proteins for a minute since the vast majority of the examples I describe in the book are proteins. First, what are proteins? In a simple sense, proteins are the main molecules in cells that makes things happen. Proteins are sometimes called the "workhorses" of the cell. Whenever something happens in a cell, you can bet that proteins are probably making it happen.

Where do proteins come from? You probably remember that the information for how to make a protein is stored in genes within the DNA, and that DNA is copied to make an RNA that serves as blueprint for the protein. A simple way to think about this is to imagine a recipe for your moms favorite dish that is written on one of those 3X5 notecards that used to be so popular. The note card is like the DNA and stores the information but you can make a copy of the recipe with a copy machine. Think of the new paper copy as the RNA. The cell uses the recipe/information in the RNA copy to make the dish. In the case of a protein, the dish is going to be a yummy chain of amino acids.

The amino acids are the ingredients for the dish and they are connected end to end according to the directions in the RNA, resulting in a protein. Sometimes, several individual proteins can be combined together to make a single larger protein with multiple subunits. Many of the proteins we will be looking at consist of four or more identical subunits. Overall, this process of DNA --> RNA --> Protein is called the "Central Dogma" of molecular biology.

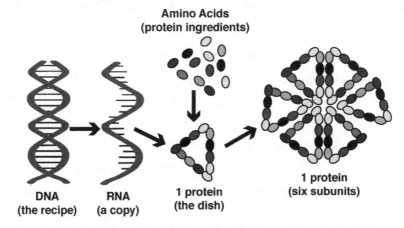

Figure 2: The Central Dogma of molecular biology.

This may be the technical description of what proteins are, but when most people hear the word "protein", images of steak and eggs appear in their head. This isn't wrong, but it's not exactly correct either. Steak is animal muscle, and that muscle is made of cells that have lots and lots of individual proteins inside. Most of the proteins in those muscle cells are "molecular machines" that are dedicated to giving muscle cells the ability to move. But these proteins are just a few kinds of proteins. We don't know the exact number, but most scientists accept that humans can make a few hundred thousand individual and different kinds of proteins. Each of these many proteins has an individual function and working together, they give cells special functions. Some proteins produce energy from the food we eat, some proteins enable cells to communicate with each other, some proteins form long fibers that give cells shapes, and some proteins allow cells and muscles to move. There is a nearly endless variety of functions that your proteins perform in your body. Understanding how these proteins function is a major part of what it means to study molecular biology.

One of the most challenging things about molecular biology is that, while there are only four basic kinds of molecules of life, there are almost limitless variations of each type. Proteins, for instance, have a mind-boggling diversity. As a wild guess, I'd have no trouble believing that there are trillions of different kinds of proteins on Earth, and this is probably a gross underestimation. No individual, no matter what their IQ is, could possibly know and understand the molecular details of all these proteins. Because of this, many of the world's most dedicated molecular biologists choose to learn everything they can about just one protein. But these molecules don't give up their secrets easily. Some people spend their entire scientific career studying how a single protein works. Through decades of research by thousands of individuals, we have gathered enough of this deep knowledge to make some generalizations about how molecules work. The good news for non-scientists is that by studying these generalizations, even an inexperienced person can get a pretty good conceptual understanding of how molecules work.

The intent of this book is simple; to help people tap into the generalizations of molecular biology that we have discovered over the years. To translate these generalizations into plain language so that we can all appreciate the elegance of molecular mechanisms. The people who I think will really benefit from this book are those who lack (but want) a conceptual understanding of biology, and their world, at a fundamental level. People who don't even know where to begin. I'm thinking of people who were interested by biology and chemistry, or even physics, in high school, but didn't have the opportunity to go to college, or maybe did go to college but majored in a different field. People who have tried to learn some molecular biology but have found it really difficult because there are no translations from deep science into common English. That's what this book is. A translation between the "geek-speak" of modern molecular biology into the common tongue we all speak in our kitchens. Or, maybe you're a person who has a biology degree and has a decent understanding of molecular biology but finds the topics of these chapter titles interesting and exciting! I'm definitively in this last category. Despite having many years of investigating and teaching these subjects under my belt, I immensely enjoyed the process of researching and writing that went into this book. I learned quite a bit myself. If you get even half as much enjoyment from reading this book as I got from writing it, you are in for a treat.

Since I am writing this book in a time where knowledge is immediately accessible on the internet, I highly recommend that you read this book with a smart phone or an internet-enabled device of some kind nearby. Most of the topics I introduce are easily searchable, and there are some real gems of videos and well-written science articles about some of them. I have pointed out a few searches that are worth your time, and I encourage you to come on this journey with me but to take whatever detours capture your interest. If you don't have an internet-enabled device – don't worry. I have tried to provide complete explanations, and my talented daughter has contributed many original illustrations that I think are both enjoyable to look at and effective at explaining difficult concepts.

As a final note before we get started, when I sat down to begin writing in the spring of 2020, during the COVID-19 pandemic, I started off by writing about electric eels. As I mentioned, eels were my initial inspiration for this project. I quickly realized that I couldn't jump right in with the eel, though, since understanding eels requires a basic understanding of topics like basic receptor-ligand interactions. I even started to worry that some readers might not remember basic chemistry or maybe what a cell is. I really struggled with balancing what introductory information to provide. I even wondered for a while if I should go way back to the beginning and discuss what atoms are. In the end I decided that, in order to keep this book short and sweet, I should assume that readers have taken, and remember at least a little, high school biology and chemistry. My goal was that anybody with that background should be able to understand and appreciate this book.

Hopefully I succeeded, and you enjoy this ride through the molecular world. By the time you reach the end, I hope you look around at your world with different eyes. I hope you come to appreciate that even the most common organisms and experiences you encounter every day are actually phenomenally elegant and amazing. I hope that you start to think about the molecular biology that is going on, in, and around you, every millisecond of every day.

Chapter 1: Penicillin

Alexander Fleming discovered penicillin on accident. That accident, along with vaccines, were arguably the most important public health discoveries of all time.

Much of what happens in molecular biology is based on the simple fact that different molecules will stick together if they have complementary shapes and chemistries. The first antibiotic, penicillin, is an excellent example of how molecular shapes define molecular interactions. The story of penicillin is also an amusing glimpse into the world of science.

Things have shapes for a reason

Take a moment to think about the world you live in. What do you see? Right now, I am sitting at the kitchen counter in my home. I see a cube-shaped refrigerator. I see flat dishes. I see a hollow coffee cup, and I see a glove. I'm sitting on a chair. I know it's silly to think about this, but **why** is the refrigerator cuboidal? So we can store food in there, and also so we can most efficiently use the space in the kitchen. Would a spherical refrigerator make for efficient space utilization? Why is the dish flat? So we can put food onto it without spilling. What would happen if plates had domed surfaces? Why is the cup hollow? So it can hold the coffee we pour into it. Why is the glove hand shaped? Because otherwise, how would it keep your hands warm in the winter? And finally, what would happen if my chair were cone shaped? I don't want to think about that one! The lesson here is that these simple things have a shape that gives them a function. **Shape dictates function**. I know it sounds elementary, and you probably already appreciate this in your everyday world, but it's the same with molecules. **Molecular shape dictates molecular function**. It's just that we are not used to thinking about the shapes that we cannot see.

Receptors and ligands: The puzzle pieces of life

All molecules have a shape that allows them to fit together with other shapes, just like your hand can fit inside of a glove, and a single, unique jigsaw puzzle piece can fill the last empty spot in a puzzle. If two molecules have complementary shapes, they can interact and stick together. When two molecules interact, they can influence each other's activities.

When we talk about two molecules interacting in biology we

use the generic words *receptor* and *ligand* to differentiate between the two molecules. Receptors are usually (but not always) larger proteins with a shape that will receive another molecule, but only if that molecule fits. Receptors are like the incomplete puzzle, receiving only the shapes that fit and are oriented correctly. The ligand is usually (but not always) a smaller molecule that binds to the larger receptor. Ligands are like the puzzle pieces, looking for a place to fit.

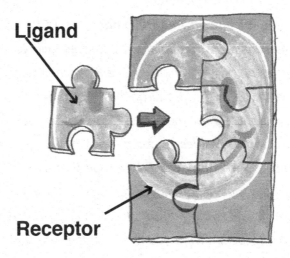

Receptors can be almost any protein, but the ones that we most often think about usually have some important function they serve inside an organism. Most receptors will either be turned on or turned off when they interact with their ligand. For instance, a receptor might be a protein that triggers the nervous system, but only when the correct neurotransmitter (a ligand) is around. Or a receptor might be a protein in your nose that can tell your brain what you are smelling, but only after it interacts with a specific ligand molecule in the air. Other examples of ligands are antibiotics, toxins, and medicines, or literally any molecule that fits a receptor like a hand in the glove. The world is filled with potential ligands just looking for a receptor.

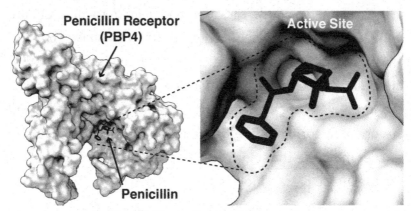

Figure 1. Penicillin (a ligand) binding to PBP4 (a receptor). *These images show the actual shape of the penicillin-binding protein 4 (PBP4) receptor with an individual penicillin molecule (ligand) in its active site. In the blow-up panel on the right the penicillin molecule is shown snugly fitting into the active site, which is outlined in a dotted line. The actual molecules are far, far too small to see in a standard microscope, so a high-power microscope that uses X-rays instead of visible light must be employed to visualize these molecules. The technique of viewing individual molecules using X-rays is called X-ray diffraction. I'll be showing you many images like this one to help you really get an understanding of how molecular shapes dictate molecular function.*

Active sites: Where the chemistry of life happens

Look at figure 1 above. It shows a small ligand tucked into the center of its larger receptor. In this case, the ligand is the original antibiotic, *penicillin*, and the receptor is a protein made by bacterial cells that binds with penicillin and is aptly named *penicillin-binding protein 4*, or PBP4 for short. Although it might look disorganized and chaotic, the receptor actually has a very specific shape that allows it to do its job. The hollow depression where the penicillin fits into the receptor is called the *active site,* and it's outlined in the enlarged panel on the right. The picture also shows penicillin tucked into the active site. See how the general size and kinked shape of penicillin allow it to fit into the active site? The active site is like the heart of the receptor. It is typically in active sites where the chemistry of life happens.

What you can't see in this picture are the chemical interactions between penicillin and PBP4 that help to "stick" penicillin into this site. Penicillin has two properties that allow it to bind into the active site on its receptor. In fact, all ligands possess these properties. First, as we have already discussed, the ligand needs to have the shape and dimensions that match the receptor's active site so that it can snugly fit into the receptor. Second, the ligand needs to have the proper chemistry that matches the chemistry in the receptor's active site.

Imagine a receptor that has a lot of negative charge in its active site. What happens if a ligand comes along that has the right shape but is also negatively charged? The ligand will be repelled and will not be able to physically get near enough to fit into the active site.

Conversely, since opposites attract, a different ligand with the same shape but a whole bunch of positive charge on it, might stick into the active site. It's the combination of proper shape and proper chemistry that allows a ligand to fully interact with the active site on its receptor.

Let's get together

Okay, so we know that ligands are small molecules that have both the proper shape and proper chemistry to fit into active sites on receptors. How does this help us understand how penicillin (or almost any other medicine in existence) works? Recall that penicillin is a medicine we take to treat bacterial infections. Penicillin kills bacteria. When a person takes penicillin, the penicillin gets into the person's blood stream and, from there, into almost every little nook and cranny in the body. All the human cells **and** the bacterial cells in the body are exposed to the penicillin. So why doesn't the penicillin hurt the human cells? **Because humans do not have a good receptor for penicillin**, therefore the penicillin has nothing to interact with and has no effect on the human cells. With bacterial cells, however, it's a different story!

The "magic" of penicillin is that its shape and chemistry are very similar to the shape and chemistry of a molecule that bacteria need for constructing their cell walls. Consider figure 2 on the next page, which compares the structure of penicillin to a component of bacterial cells walls called *acyl-D-alanyl-D-alanine*, which we will abbreviate to ADADA. What differences and similarities can you identify?

The basic structures of ADADA and penicillin are roughly the same and some of the chemistry is also the same. Critically, some of the chemistry is also very different. In fact, you might consider penicillin to be a "Trojan horse". It appears safe at first but hides a nasty chemical secret that can kill bacterial cells.

Normally ADADA binds to its receptor, PBP4, where it is combined with other molecules to form the bacterial cell wall, which is required for the bacterial cell to live. The shape and chemistry of penicillin is so close to the shape and chemistry of ADADA that penicillin can also bind into the same active site on PBP4. When it does this, it is competing with ADADA for the use of the receptor. Competition between a natural ligand and an artificial or medicinal ligand for the same active site is a common way that many medicines work. By itself, competition between penicillin and ADADA would probably slow bacterial growth, but penicillin also has another way to disable PBP4.

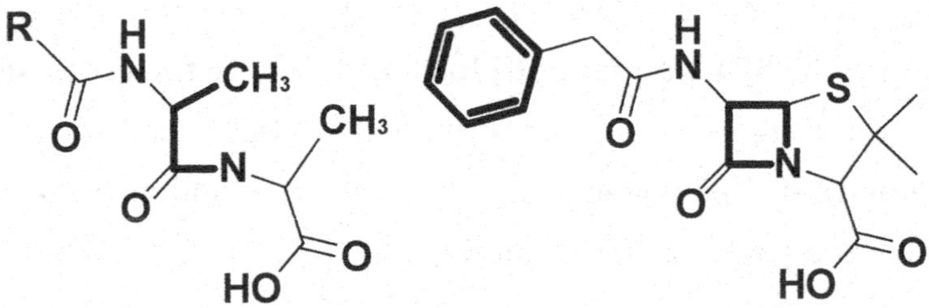

acyl-D-alanyl-D-alanine (ADADA)   **Penicillin**

Figure 2: Comparison of the chemical structures of acyl-D-alanyl-alanine (ADADA) and penicillin. Pay particular attention to the chemical bonds indicated with heavy lines. In penicillin, the heavy square shape is called the lactam ring. The lactam ring is a characteristic that all lactam antibiotics like ampicillin, penicillin, and amoxicillin share. ADADA has equivalent chemical bonds, indicated by heavy lines, but notice that in ADADA the bonds do not form a full square. Note, too, that in ADADA the "R" indicates a longer chemical structure attached at that position but not completely shown in the figure. In fact, ADADA is just the end of a long protein that is used to make the bacterial cell wall. On penicillin, instead of an R, there is a bold phenolic ring. Penicillin is a tiny individual molecule, it is not attached to a larger molecule like ADADA is.

Figure 3 is the same picture as in figure 1; except this time, I have used a *ribbon model* instead of a *space filling model* to illustrate the receptor and its active site. The ribbon model lets us look at the protein's inner structure and examine individual parts of the receptor. The arrows and helices indicate the structures that the amino acid chain is folded into. Using the ribbon model, we can zoom in to the active site and actually see that a new chemical bond is formed between penicillin and the receptor. This happens because penicillin's chemistry is slightly different than ADADA's, so instead of being incorporated into the growing bacterial cell wall, penicillin forms a new chemical bond with PBP4, **permanently locking itself into the active site**. In the picture, the new chemical bond has been formed where the black-shaded ligand is attached to one amino acid sticking out of the gray-shaded receptor.

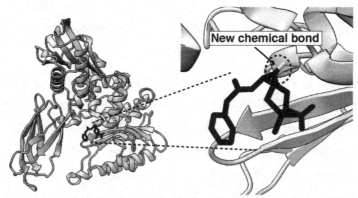

Figure 3: Ribbon model of penicillin receptor with penicillin binding. *On the left is the whole penicillin receptor with penicillin in the active site. On the right is a blow up of the active site showing the new chemical bond that has formed between penicillin and the receptor protein. Once this chemical bond is formed, penicillin cannot be removed from the receptor, ADADA can no longer bind, and the receptor protein is no longer useful to bacterial cells.*

Once the active site is blocked by penicillin, ADADA cannot gain access to PBP4. The receptor is effectively dead. I guess you could call penicillin sort of a chemical kamikaze! A chemikazi? If all the bacterial ADADA receptors are deactivated by penicillin sticking to their active sites, the bacterial cells **cannot make cell**

walls. Bacteria without cell walls undergo a process called *lysis* where they basically swell and pop like little balloons that have been overfilled.

Let's summarize the key facts that make penicillin effective against bacterial cells but harmless to humans. First, while the bacterial cells do have a receptor protein for penicillin, human cells do not, so penicillin specifically interacts with bacterial cells but passes harmlessly by human cells. Second, the structure of penicillin mimics the natural ADADA molecule, allowing penicillin to enter into the active site of PBP4. Third, the kamikaze chemistry of penicillin allows it to form a new chemical bond with PBP4, permanently blocking ADADA from binding to the active site and preventing cell wall synthesis.

This example serves as a simple introduction to how molecules work. Successful interaction between any ligand and its receptor, or really any two molecules, almost always depends on compatible shapes and chemistries between the interacting molecules. This concept will be applied immediately in the next chapters when we discuss and compare toxins and medicines, but it can also be applied to almost any molecular interaction in biology.

The story of penicillin is not over. Before we leave this topic, here are a couple more scientific facts about the discovery of penicillin and the emergence of antibiotic resistant bacterial cells that will help to further illustrate how molecular shapes and chemistries can be slightly altered to change molecular functions.

The accident of penicillin: Luck favors the prepared

The discovery of penicillin is one of the greatest stories of accidental scientific discovery ever told. In 1928, Dr. Alexander Fleming was growing common *pathogenic* (disease causing) bacterial cells on bacterial growth plates and discovered, by accident, that a fungus was also growing on his plates. Instead of just throwing the plates away, Dr. Fleming carefully observed the plates and noticed that bacterial cells were unable to grow near the fungal cells. Figure 4 on the next page is an actual photograph from the 1929 publication showing the original growth plate where Dr. Fleming made this observation [2]. In his notes, Dr. Fleming indicates the accidental *colony* (cluster of cells) of the *Penicillium* fungus and normal *Staphylococcal* bacterial colonies growing at the bottom of the plate. *Staphylococcal* bacterial cells are common human pathogens that cause diseases such as skin boils, scalded skin syndrome, and toxic shock syndrome. In the middle, Dr. Fleming indicates "Staphylococci under-going lysis", pointing out *Staphylococcal* cells that are being negatively affected by the *Penicillium* colony. From this observation, Dr. Fleming hypothesized that the lysis was caused by molecules being secreted by the fungus. He later managed to isolate the molecules and named it penicillin, since it was isolated from the Penicillium fungus. This simple observation led to one of the most important discoveries in human history!

Some might call this accidental discovery "luck", and there is certainly some of that in the discovery of penicillin, and in all science for that matter. After all, it was very lucky that Dr. Flemings' experiment was contaminated with a Penicillium colony instead of some other fungus. However, without his curiosity, perception, and patience, Dr. Fleming probably would have just thrown this contaminated plate away. Instead, he recognized this accidental discovery for what it was: a powerful tool against bacterial cells, which, in 1929, were one of the leading causes of death in the world. Back then, a simple pimple on your nose could possibly kill you! Starting with Dr. Fleming's observation, the development of antibiotics has easily been one of the most important, if not **the** most important, public health discoveries of all time.

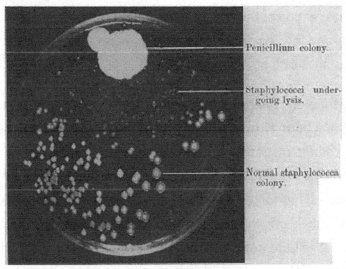

Figure 4: A bacterial growth plate showing the antibiotic activity of penicillin fungus. *This picture is taken from Dr. Fleming's 1929 publication where he first announced to the world the discovery of penicillin.*

An ancient arms race: On the front lines of molecular warfare

Dr. Fleming was quoted as saying "I did not invent penicillin. Nature did that. I only discovered it by accident." After reading this, you might be wondering, "why does this fungus make penicillin in the first place?" As it turns out, there has been an ongoing war between bacteria, fungi, and other organisms in the world since life first began on Earth. Many of these organisms live in the same environments, compete for the same resources, and have evolved weapons to help them kill their competitors. Sound familiar? Scientists have found that many fungi and even some bacterial cells produce antibiotic molecules, and we have harnessed some of these to create new antibiotics for human use. These molecules originally evolved slowly, over time, in response to a biological need to "fight back" against microbial competitors. Amazingly, the evolutionary forces that allowed for the creation of these antibiotics continue to shape our world today, and the development of antibiotic resistance in bacterial cells is an excellent example of this.

Unfortunately, the misuse of antibiotics for things like treating viral infections (viruses do not have the correct receptors so they cannot respond to these antibiotics) or overuse in the agricultural world (continually exposing bacteria to these molecules) has forced bacterial cells to "fight back" by evolving new mechanisms to resist the action of antibiotics. Of course, bacterial cells do not have intelligence; they do not make these choices on their own. Instead, bacteria simply allow genetic variation within their populations and evolutionary forces to shape their response to human antibiotics. Bacterial cells are able to rapidly evolve like this because they reproduce very quickly, and with every new bacterial cell that is made comes another chance to evolve new traits.

In the case of penicillin, an enzyme named β-*lactamase* has become prevalent in bacterial populations. β-lactamase physically cuts, or severs, chemical bonds in penicillin as shown in the figure on the next page. Once it is cut, penicillin no longer has the proper shape or chemistry to fit into the active site of the receptor, so it cannot interact with the receptor. Free,

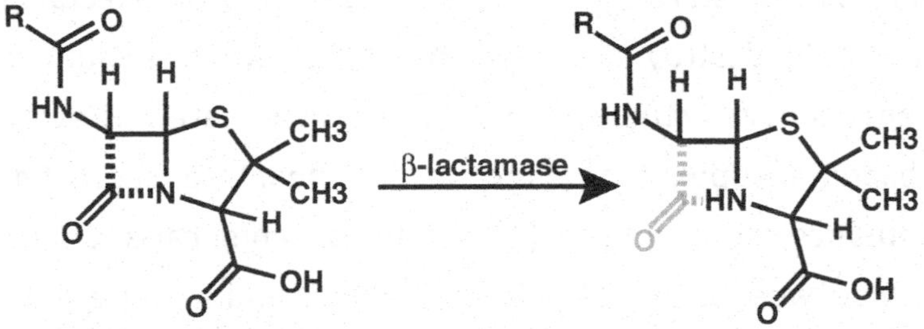

Penicillin | Inactivated Penicillin

Figure 5. ***β-lactamase function.*** *The molecular bonds indicated by dashed lines are cut by the enzyme beta-lactamase, destroying the lactam ring. This one simple change in the shape of penicillin inactivates it so that it cannot interact with the ADADA receptor.*

non-penicillin-bound receptors allow ADADA to continue being incorporated into cell walls, ensuring bacterial survival.

Since penicillin is a naturally occurring antibiotic, β-lactamase has probably been present in bacterial populations for almost as long as fungi have been making penicillin. This is how evolution works. Natural gene variants in a population may provide an advantage to individuals that make that gene. Charles Darwin would call this "survival of the fittest". In the case of penicillin, human overuse and misuse of penicillin forced a selective pressure onto bacterial populations, which led to the survival and proliferation of bacterial cells that have the β-lactamase gene. All of this has contributed to the rise in recent years of bacteria that produce β-lactamase proteins and no longer respond to penicillin treatment.

As bacteria developed antibiotic resistance, humans needed new weapons. We created new antibiotics that acted like penicillin but had chemical modifications that made them resistant to β-lactamase. Of course, the bacterial cells fired back, evolved, and started making new variations of β-lactamase that could destroy the new antibiotics. Some evolutionary scientists have actually studied how β-lactamase has evolved in bacterial populations in response to the introduction of new antibiotics over the years [3]. It's a fascinating cat-and-mouse story between bacterial cells and humans and serves as a very relevant reminder that evolutionary forces are still at work even in the 21st century.

If you want to see a fascinating example of bacterial evolution, in 2016 a group of scientists devised a way to directly observe the evolution of bacterial antibiotic resistance [4]. In their experiment, bacterial cells evolved resistance to very high levels of antibiotics in about eleven days! I strongly suggest you do an internet search for a video of "evolution of bacterial cells on a mega plate" or something to that effect. It's one of the best examples of evolution-before-your-eyes that I have ever seen.

Eventually, humans made a super-weapon – a new antibiotic, *methicillin*, that was structurally and chemically similar to penicillin but was resistant to destruction by all β-lactamase enzymes. Discovered in 1960, methicillin was, in its day, the best antibiotic humans possessed in their arsenal of weapons against bacterial pathogens. Alas, as the war has waged on, methicillin-resistant bacterial cells have evolved.

One of the most serious antibiotic resistant organisms currently threatening humans is called *methicillin-resistant Staphylococcus aureus*, or MRSA for short. Ironicly, MRSA is a variant of the original *Staphylococcus aureus* microbe that Alexander Fleming first used to discover penicillin. To defeat our super-weapon, methicillin, MRSA uses its own "secret weapon"

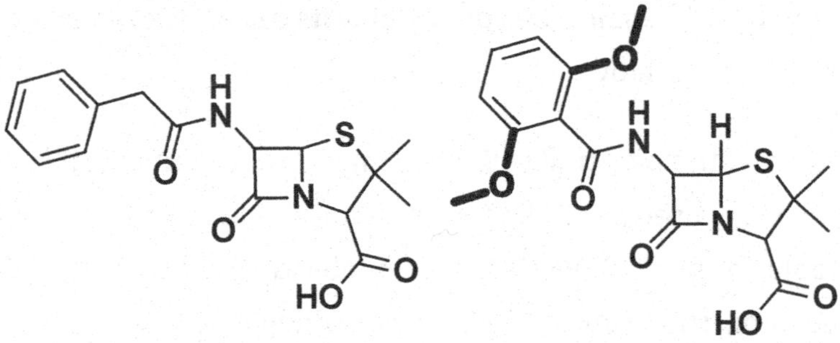

Penicillin   **Methicillin**

Figure 6. Penicillin vs methicillin. Methicillin has the same lactam ring and a molecular structure very similar to penicillin's. Its shape and chemistry allow it to bind into the ADADA receptor's active site as we discussed above with penicillin. A few minor changes, shown in bold lines, prevent the molecule from being cut by β-lactamase, making it a useful weapon against the bacteria that can easily destroy penicillin.

from the bacterial cell arsenal. Many bacterial cells, including *Staphylococcus aureus,* have a trick where they can pick up bits of DNA from the environment and try them out. If the bit of DNA provides an advantage, such as antibiotic resistance, that DNA can be kept and passed to future generations, thus spawning a wave of new antibiotic resistant bacterial cells. In the case of MRSA, bacterial resistance is linked, not to new variations in the β-lactamase gene, but rather to the gene for a protein called *penicillin-binding protein 2a*, or PBP2a [5]. The PBP2a protein is another protein involved in building cell walls that can **replace the normal PBP4 protein**. The new PBP2a protein has an active site that also recognizes and can use ADADA to build walls. The catch is that PBP2a has just a little different shape and chemistry than the normal PBP4 receptor. The slightly altered shape and chemistry change PBP2a's active site such that **methicillin can no longer bind into it**. Methicillin cannot compete for the active site, and MRSA happily goes on building its cell walls even when methicillin is close by.

So, you see, the bacterial cells have managed to "outsmart" us. We originally used penicillin to kill bacterial cells since penicillin was able to mimic ADADA and compete for the active site of its receptor. The bacterial cells responded with a weapon that destroyed penicillin. In retaliation, we created a new antibiotic that could not be destroyed by the bacterial weapon. Finally, the bacterial cells have responded by changing the active site of the receptor, which essentially takes us back to the beginning. At this point, the bacterial cells are winning the war and many of our antibiotics are no longer useful for treating infections.

It all comes back to shapes and chemistries. Molecules must have the correct shapes and the correct chemistries to be able to interact with one another. These shapes and chemistries are developed naturally over time in response to selective pressures, and, sometimes, they can be used in tricky ways to give one organism an advantage over another.

As we move forward and discuss molecules with really cool functions it will be important to keep the terminology straight; these terms are our building blocks. Receptors are biological proteins that have an active site to which a ligand can bind. If both molecules have the right shapes and the right chemistries, interaction can occur. Although they might seem trivial, the interactions between receptors and ligands are indispensable for life on Earth.

Chapter 2: Toxin or Medicine?

Opossums might not be pretty, but they have a pretty neat trick that makes them resistant to rattlesnake bites. Opossums have ligands in their blood that bind to, and neutralize, the proteins that make rattlesnake venom deadly.

As I was doing research for this book, I wanted to gauge the general level of understanding that people have about the molecular world around them. I asked people this simple question, what is the difference between a toxin and a medicine? Most people responded with something to the effect of "medicines cure and toxins kill." That answer is technically correct, but it also indicates that most people don't have a deeper understanding of **how** molecules work to impact our everyday lives. And that's okay. Most people never need to think about how molecules work. However, without this basic understanding, things like medicines and toxins begin to take on a mysterious and magical nature. Arthur C. Clark was the famous science fiction writer who wrote the phrase, "Any sufficiently advanced technology is indistinguishable from magic." The goal of this chapter is to reinforce the idea that medicines and toxins are not magic. The truth is much simpler. Medicines and toxins are just molecules that, much like penicillin, function by interacting with and either increasing or decreasing the function of specific receptor proteins. No magic required, just science!

Toxins versus medicines: A matter of dose

So what is the difference between a toxin and a medicine? As it turns out, toxins and medicines are far more similar than they are different. Both toxins and medicines are just molecules that interact with receptors in your body. Whether they hurt or heal often depends on what receptor the molecule binds to and/ or how much of the medicine or toxin is given. For instance, if a receptor has an important role in your body, an abnormal molecule binding to it might make it malfunction and make you sick. We would probably call this a toxin. On the other hand, if you have a disease that is caused by an overactive receptor protein, perhaps binding a molecule to it will settle it down and be beneficial. We would probably call this a medicine. In both cases, the toxin and the medicine are fundamentally doing the same thing. They are just binding to, and changing the function of, a receptor. Sometimes a toxin and a medicine might actually bind to the same receptor but in different ways. In many cases, the only difference between a medicine and a toxin is dose. Actually, some of the most potent toxins we know of work as medicines if given at low doses (think Botox). At the end of the day, toxins and medicines are just molecules that interact with receptors. Let's talk about some specific examples to better understand what makes a molecule a toxin or a medicine.

Snake venoms: Taking control of the cardiovascular system

One snake that should really get your attention is the Russell's viper. This beautiful snake is found in parts of Asia and is a major threat to people who work outdoors in that part of the world. Its venom is very deadly to those bitten. This snake first got my attention when I watched a video showing what happened to a cup of blood after adding a single drop of Russell's viper venom. Within seconds that cup of blood jellified into a large blood clot that looked like a squishy, crimson-colored hockey puck. Take a moment to look around the internet and find a video that shows the effect of Russell's viper venom on blood. It's stunning, amazing, and terrifying all at the same time!

Russell's Viper

This rapid blood coagulation is how the Russell's viper kills its prey; it causes blood clots to form in the blood stream, and those clots block blood flow to essential organs. Abnormal coagulation of blood is a very serious medical emergency and is involved in two of the most common causes of death: strokes (blood clots in the brain) and heart attacks (blood clots in the blood vessels of the heart).

In the 1950s and 1960s, scientists were working to understand how blood clotting happens in humans so they could address the problem of strokes and heart attacks. When you have very little knowledge about a topic, sometimes it helps to find a good model system to study. A model system in biology is an organism that has properties you are interested in studying. As an example, mice are a model system for human disease since they are also mammals and get many of the same diseases that humans get. In the case of blood clotting, it had long been known that the Russell's viper has a profound effect on blood clotting, so this seemed like a good model to use to study blood clotting.

In order to dispel the "black magic" of blood clotting, the scientists needed to know about the molecular interactions that cause blood clotting. To learn about these interactions, they needed tools. Scientists reasoned that somewhere in the viper's venom lay some hidden molecule that might give them a clue as to how blood clotting works. If they could just discover this molecule, they could use it as a tool to answer questions about how blood clotting normally happens. So, scientists set out to study the components of the Russell's viper venom. After many experiments, *Russel's viper venom 5* (RVV-5) and *Russel's viper venom 10* (RVV-10) emerged as the two molecules that were primarily responsible for the viper's "black magic" of rapid blood clotting.

The molecular biology of blood clotting: Dams, dominos, and hand grenades

In order to understand how these snake toxins work, I'm going to need to take you on a little detour into the world of blood clotting. Once you understand a little of how the blood clotting systems work, you will then be able to better understand how the snake venoms take advantage of this system to cause pathological blood clotting.

Blood clotting is when you get a cut but it doesn't bleed for very long because the wound seals itself up. The seal that physically stops the bleeding is the blood clot. I suspect you have experienced this yourself, perhaps many times, but what exactly is happening at the molecular level to form that blood clot? It's much more than just blood drying up and hardening like glue. To imagine what's going on during blood clotting, think of the blood flowing through your arteries and veins as water flowing down a river. In a river, the water and all the debris, sticks, leaves, and everything else, all float downstream just fine. But sometimes something like a big tree might fall into the river. Or a big log might get stuck across the river. When this happens, all those sticks and leaves start to pile up behind the fallen tree and form a natural dam that causes the water to back up behind it.

In this analogy, the "fallen tree" is the equivalent of a blood protein called *fibrin.* Fibrin is the main structural core of the blood clot. The "sticks and leaves" that collect behind it are blood cells and platelets that get stuck in the fibrin. Eventually the "debris" fills in all the holes and makes a leak-proof clot. Here's the important question: what causes the tree to fall (or the fibrin to start the clot) in the first place? It's actually a molecular chain reaction that begins with damage to blood vessels and ends with fibrin starting a clot. Understanding this chain reaction is the molecular key to understanding how the Russell's viper and some other snakes use their venoms to kill.

This chain reaction of protein activation and blood clotting is called the *clotting cascade,* and it might appear to work sort of like a Rube Goldberg machine. If you don't know what a Rube Goldberg machine is, these are the overly complex machines that have dozens of intricate and unnecessary steps just to do something simple like put some toothpaste on your toothbrush. If you are not familiar with Rube Goldberg machines, the band

"Ok Go" has an exceptional example in the video for the song "This too shall pass". It's really fun and worth a watch. Just like all Rube Goldberg Machines, the clotting cascade is started by a small push that sets off a complex chain of events that ultimately end with fibrin and a blood clot.

The clotting cascade is normally activated by damage to blood vessels, either when blood leaks from a vessel, or when the cells lining a vessel become damaged. There are certain molecules outside the blood vessels that blood would never normally interact with. Once a blood vessel is broken, however, the blood mixes with these molecules. The molecules from outside can now interact with proteins in the blood called *clotting factors,* and these interactions kick off the clotting cascade. After the first push, the clotting cascade progresses through a series of sequential steps where one clotting factor activates the next, which activates the next, which activates the next, just like one falling domino makes the next domino in line fall down. With the domino analogy in mind, imagine that the dominos are arranged in a pyramid formation; when you push the first domino at the apex of the pyramid, it knocks down two dominos, which knock down three more, then four, then, five, and so on. The cascade gets stronger and stronger until dozens of dominos have fallen in a short period of time. This is the first part of the Rube Goldberg blood clotting machine.

Clotting factors are nothing more than proteins floating around in the blood. What makes them special is that they have an extra bit of protein attached that needs to be cut off before they can be turned on. This extra bit of protein locks the clotting factor in an inactive state until it is needed. The extra bit of protein works as if it were a ligand stuck in the clotting factor's active site; once it's cut off, the active site is free. The extra bit is a kind of safety mechanism that almost all clotting factors have. Removing it is sort of like pulling the pin from a grenade. Once the "pin" is pulled out, the clotting factor becomes an active enzyme. It can then "pull the pin" from, and activate, the next clotting factor in line. Once the clotting cascade is started, the "dominos" (clotting factors) start falling faster and faster, and this eventually leads to the activation of a clotting factor called *factor 10*. Once the inhibitory "pin" of factor 10 is removed, its name changes to *factor 10a* ("a" for "activated"), and this is when the clotting cascade really kicks into high gear.

Factor 10a activates *prothrombin*, again by removing a bit of protein, or "pulling the pin." Prothrombin is now called *thrombin*, and it has two important jobs. First, thrombin activates *factor 5*, which helps to activate even more thrombin, forming a *feed forward loop*. Second, thrombin cuts a protein called *fibrinogen* to make fibrin, which, back to our original river analogy, acts like the tree falling across the river.

This last step of fibrinogen getting cut to fibrin is very interesting and deserves even more attention. Fibrinogen easily dissolves into your blood and normally floats through your blood stream along with everything else. When thrombin cuts fibrinogen in the last step of the clotting cascade, the resulting fibrin is no longer able to dissolve into blood. Instead, fibrin begins to precipitate near the wound site forming an insoluble protein meshwork that traps red blood cells, white blood cells, and platelets into it. If you do an internet search for "blood clot SEM", you will find lots of <u>s</u>canning <u>e</u>lectron <u>m</u>icrograph (SEM) pictures of what these blood clots actually look like. They remind me of a box of ropes where all the ropes are haphazardly intertwined with each

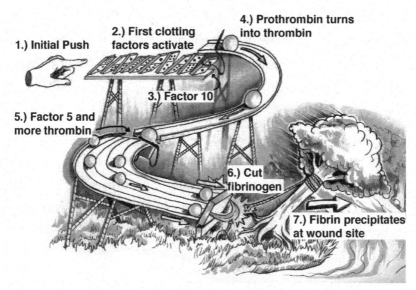

1.) Initial Push

2.) First clotting factors activate

4.) Prothrombin turns into thrombin

3.) Factor 10

5.) Factor 5 and more thrombin

6.) Cut fibrinogen

7.) Fibrin precipitates at wound site

Figure 1: Blood clotting cascade as Rube-Goldberg machine.
Blood clotting is given the "first push" when blood vessels break and the first clotting factors activate each other like falling dominos. These first clotting factors activate factor 10 which converts prothrombin to thrombin. Thrombin activates factor 5 which activates even more thrombin. Finally, thrombin cuts fibrinogen to make fibrin which precipitates in the wound and seals the blood vessel to stop bleeding.

other in a hopeless knot. The result is a mass of sticky proteins and cells that forms a dam and stops blood flow. Think of how that falling tree causes a logjam in the river and all the floating stuff piles up behind it, eventually making a natural dam on the river. This is basically the same thing that happens when fibrin precipitates; in blood clotting, fibrin molecules are the logs that rapidly plug up the hole to stop the bleeding. Amazingly, this whole process normally happens in only a few seconds.

I just described to you the basic steps of blood clotting, but we need to pause for a moment to appreciate this system. There is quite a bit of elegance in blood clotting that might not be readily apparent! First, the interactions between clotting factors are nothing more than a series of receptor-ligand interactions. Second, all of these proteins are enzymes, meaning they catalyze a chemical reaction, and they will continue their work until they are specifically turned off. That means that each activated clotting factor can activate many more factors, which activate many, many more factors, which activate many, many, many more factors, and so on in a chain reaction. Other examples of chain reactions include forest fires, the explosion of atomic bombs, and the rapid spread of viruses like SARS-COV-2 across the planet. All pretty violent things.

Another reason I find the clotting cascade so elegant is that these clotting proteins are in your blood all the time but, like the grenade, need to be activated in order to work. An important advantage of this strategy is that it allows organisms to store large amounts of clotting factors in an inactive, but ready, state so that they can respond to damage within seconds. If these proteins were not always present in your blood it would take your body a long time, maybe hours, to produce enough clotting factors to stop a wound from bleeding, during which time you could lose a lot of blood, even from a small wound. It would be sort of like having to first set up your Rube Goldberg machine before starting it up. Luckily, your body already has the dominos, balls, axes and all the other components of your Rube machine set up, so it just takes one push to set off the chain of events.

Before we move on, what do you think would happen if a single component of the Rube Goldberg machine failed? The gap between steps would stop the chain reaction, right? This is actually what happens in many forms of *hemophilia*, a clotting disorder where minor injuries like cuts and bruises could end up being fatal. Perhaps the most well-known hemophiliac of all time was Alexiei Romanov, the son of the last Russian Czar, Nicholas Romanov II. Alexiei suffered tremendously as a child. In those days, the early 1900s, hemophilia was still a mystery and could only be treated as such. Rasputin, the infamous Russian mystic and healer, treated Alexiei, relying on unproven treatments and supernatural faith healing. Today we understand how the clotting cascade works. We know that if you take away any one step in the Rube Goldberg device, the chain of events will fail and blood clotting will either be slow or completely absent, depending on which clotting factor is missing. In Alexiei's case, he lacked clotting factor 9 [6]. With knowledge of the molecular interactions required for normal blood clotting, we can now efficiently treat the disease by simply injecting pure clotting factor 9 when it's needed. Hemophilia no longer has to be an immediate death sentence!

Russell's viper venom: Covert jellification

So how do the Russell's viper and its venom fit into this picture of blood clotting? The answer is hidden in the names of the venom proteins. RVV-5 and RVV-10 are named for what they do; they artificially activate clotting factors 5 and 10 and trigger abnormal clotting. RVV-5 and RVV-10 **bypass the entire first part of the cascade and directly trim the inhibitory bits off of factor 5 and factor 10**. They cheat the Rube Goldberg machine! It's like pushing the grand finale cannonball onto the toothpaste tube yourself without letting all the preliminary machines do their jobs. It's a short circuit.

A critical difference between the natural system and viper venom activation is that natural blood clotting is strictly restricted to the location where a blood vessel was initially damaged. In contrast, RVV-5 and RVV-10 are injected into the blood stream and go everywhere your blood goes, which is essentially everywhere in your body. The result is that, as RVV-5 and RVV-10 float through the bloodstream, they form blood clots that float around and get stuck in blood vessels all over your body. Remember that strokes and heart attacks are two of the most common causes of death in humans, and that these diseases are often caused by a single abnormal blood clot in the circulatory system. Imagine the damage that could be caused by blood clots throughout a person's entire bloodstream! It's not difficult to understand why a bite from the Russell's viper is so dangerous.

Why study snake venoms?

People have been studying snake toxins for decades. Why? Not because they have a morbid fascination with death, but because the very fact that snake venoms are deadly means that these molecules are probably powerful ligands that interact with important receptors in our bodies. In the early days of studying blood clotting, scientists knew that blood could clot, but they didn't know how. They also knew that snake venom caused clotting, but they didn't know why. When scientists tried to figure out how Russell's viper venom caused blood clotting, they discovered RVV-5 and RVV-10 and this gave them a new tool to help understand how blood clotting works. They rationalized that RVV-5 and RVV-10 were probably interacting with another protein in the blood that was important for blood clotting. A clotting factor. So, they looked for blood proteins that specifically stuck to RVV-5 and RVV-10 and eventually discovered clotting factor 5 and clotting factor 10.

In these studies, it didn't matter that RVV-5 and RVV-10 were toxins. All that mattered was that they had biological activity, meaning that they probably interacted with something important (a receptor). The same thinking can be applied to any toxin. All toxins are possible ligands for biologically important receptors and can be used as scientific tools to help discover and understand how our various systems work.

Drug development companies spend millions of dollars every year searching through "drug libraries" of thousands of random molecules that might be ligands for various receptors. This method of finding ligands for receptors is a haphazard, unpredictable, expensive, and time-consuming way to find biologically active molecules. Another way to do it is to study the natural world, where Mother Nature has already given us major hints about some important ligands that fit important receptors. Like toxins. Discovering how a ligand works, even if it can't be adapted into a medicine, almost always gives us a clue about how our most important systems work and sometimes, a new tool to study these systems.

Scientist's research of snake venoms have lead to the discovery of several molecules that affect the cardiovascular system. Since problems with the cardiovascular system are at the top of the list for the most common causes of death in humans, you can understand the importance of studying toxins that affect it. For example, one Brazilian pit viper (*Bothrops jararaca*) injects a toxin that causes rapid loss of blood pressure by simultaneously dilating blood vessels and blocking the synthesis of a hormone called *angiotensin*, which helps to maintain blood pressure. This double whammy on blood pressure regulation causes bitten animals to rapidly pass out and become an easy meal for the viper. Studying this toxin in the 1980s lead to the development of a new medicine, *captopril*, and eventually, a whole new class of drugs, called ACE inhibitors that are used to control high blood pressure in people [7].

This was the first example of a snake toxin being used as a medicine, but there have been many more [8]. Another example is a toxin called *batroxobin* which is found in the venom of the Brazilian pit vipers *Bothrops atrox* and *Bothrops moojeni*. Batroxobin is deadly because it destroys fibrinogen in the blood stream in a way that prevents it from clotting. With reduced levels of clottable fibrinogen, the vipers' victims can't form clots, and, again, they bleed out and become easy meals. Studying batroxobin allowed scientists to eventually design a new drug called *Defibrase* that is used to reduce blood clotting in people who have elevated risk of strokes. Examples like these go on and on.

So why study snake venoms? Because snake venoms contain ligands that bind to receptors in the cardiovascular system and change the activity of these receptors. **This is basically the same thing that some medicines do.** In other words, snake venoms can, in some cases, be medicines. It doesn't matter whether a ligand comes from a snake, a plant, the dirt, or from outer space; all ligands are just molecules that fit into receptors and can sometimes change how the receptors work. If the receptor does something important, the ligand might have a "biological activity" and be useful to scientists or doctors. In the case of snake venoms, the difference between healing and harming is not whether or not there is a receptor, as in the case of penicillin, but in **precisely controlling the amount** of the ligand that is available to the receptor. If you can do this, a toxin can become a medicine.

Russell's Viper

Intermission 1: Neurons and Nerves

Some of the most interesting toxins are neurotoxins that work by incapacitating the nervous system. To appreciate the neurotoxins I want to tell you about, you'll first need to have a little understanding of how nerves work. This is a complex topic, but I'm going to break it down to the most basic information needed to really appreciate the elegant molecular biology here.

Nerves are really amazing. You probably know that your nerves carry messages back and forth through your body, but did you ever stop to wonder exactly **how** they do this? Or even what the heck those mysterious "messages" are made of? It's funny how we take these things for granted, isn't it?

Nerves are made up of many individual, long neurons. Neurons are "electric" cells that can sense or receive a stimulation, turn that stimulation into an electrical signal, transmit the electrical signal down the length of their bodies, and then relay this signal to the next cell. The next cell is sometimes another neuron, sometimes a muscle, or even sometimes an organ. The human body easily has trillions of individual neurons, and these cells can transmit electrical signals at speeds of up to two hundred miles per hour through your body!

If you are asking yourself "how can neurons be electric?" I'll describe what this means in chapter 4 when I also describe how electric eels work. Both neurons and electric eels use similar systems to generate electricity. For now, it's only important to understand that neurons use electricity to move signals down the neuron.

The picture below shows a typical cartoon of an individual neuron. There are four basic parts to most neurons including the dendrites, cell body, axon, and axon terminals. Dendrites are often the part of the neuron that receives signals from other neurons or parts of your body. The cell body is the central part of the neuron where the nucleus and most of the other cell organelles are stored. The axon is a long extension, sometimes coated with a layer of insulating myelin, that transmits the signal some distance. Finally, the axon terminals are the sites where the message gets delivered to the next cell.

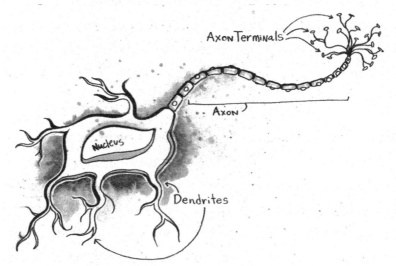

Figure 1: Diagram of a neuron.

To be precise, neurons are not nerves. Individual neurons get bundled together into the rope-like and insulation-wrapped bundles that we call nerves. It's impossible to be precise, but there are estimated to be about 100 billion nerves in the human brain and trillions of nerves in the human body. Of these, we have mainly characterized the larger nerves of central importance. The largest nerve in the human body is the *sciatic nerve,* which runs all the way from your lower back to your ankles. At its thickest, the sciatic nerve is nearly an inch thick, and depending on how tall a person is, can be up to, or even more than a meter long! Bioelectrical signals are passed all the way down this nerve at lightning speeds. Let's take a look at how these electrical messages actually work.

Separation of charges: Setting up for success

There is one very important property of neurons that we need to cover, and understanding it is crucial to understanding neural messages. Neurons are *polarized*. This means that neurons have more negative charge inside the cell and more positive charge outside the cell. The charge imbalance in a polarized nerve is caused by an unequal inside-versus-outside distribution

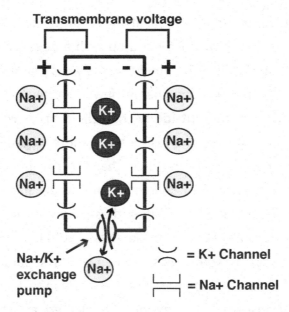

Figure 2: A polarized cell. Resting cells are polarized, meaning they are negatively charged inside and positively charged outside. This separation of charges, or trans-membrane voltage, is created by different numbers of Na+ and K+ ions inside and outside the cell. Separation of Na+ and K+ is achieved by Na+/K+ exchange pumps that move Na+ out of the cell and K+ into the cell.

of sodium, potassium, and calcium ions. Your high school chemistry teacher probably taught you that any atom with an imbalance between its number of protons and electrons will have a charge, either positive (more protons) or negative (more electrons), and we call that charged atom an ion. The potassium (K^+), sodium (Na^+), and calcium (Ca^{++}) ions **all carry positive charges**. Overall, the outside of a neuron has more sodium and calcium compared to the inside, while the inside has more potassium than the outside. This unequal distribution of sodium and potassium is established by a special Na^+/K^+ pump protein that moves Na^+ ions out of the cells and simultaneously moves K^+ ions into the cell.

Here is an important question, if all of these particles are positively charged, how do we end up with a negative charge on the inside of the cell? The simplified answer is that the inside of the cell has **less positive charge** (fewer positively charged ions) **compared to the outside of the cell**. In other words, **in relation to the outside, the inside has a more negative** (less positive) **charge.** The end result is that the cell is polarized by unequal distribution of charge on the inside and outside with the cell membrane separating these charges. Whenever you have a separation of charge like this, a voltage is created. Actually, voltage is sometimes defined as "a separation of charge". When the cell membrane is separating these charges, we call it a *transmembrane voltage*. In a cell there are about 70 millivolts, or 0.07 volts, of charge across the cell membrane. In comparison, a typical AA battery has 1.5 volts, and the standard electrical outlet in your home has 120 volts. Compared to these, cellular voltage is miniscule, but this voltage is critical for nerve function.

The polarization of a nerve is critical because nerves send messages by depolarizing. You can think of depolarization as basically the reverse of polarization. To polarize the cell, we had to move sodium and calcium ions out while moving potassium ions into the cell. The first thing that has to happen in order to make a neuron depolarize, or fire, is the opposite; the cell allows positively charged Na^+ and/or Ca^{++} ions to enter the cell. This reduces the amount of polarization, meaning it reduces the separation of charges, meaning it reduces the voltage. This first step is the most important! All the other steps happen in response to positively charged ions entering the cell. It's another chain reaction.

It's not easy to get ions into the cell though, because every cell on Earth has a cell membrane surrounding it, and these membranes make an extremely effective barrier to ions. That's one of the cell membrane's most important jobs, to separate the cell into an outside and an inside, and it's absolutely critical for life as we know it. Nonetheless, in order for the cell to depolarize, the ions still need to move into and out of the cell in a controlled way and for this, we need channel proteins.

Channel proteins: The gateways to the nervous system

The universal solution to the problem of moving ions in and out of the cells is to make holes in the cell membrane. But the holes cannot just be holes! Non-selective holes would allow ions to move back and forth across the membrane willy-nilly and would actually kill the cell. In fact, one of the ways that your immune system kills bacterial cells is by poking large holes in the bacterial cell membrane, which causes bacterial cells to uncontrollably and completely depolarize. In neurons, instead of large killer holes, dual-function proteins are embedded in the cell membrane. These proteins simultaneously serve as receptors for various ligands and as holes that open and close in response to those stimuli. Calling these proteins "hole proteins" is not very accurate. Instead, let's call them *channel proteins,* or *ion channels*, if you prefer. These terms are interchangeable.

If you can imagine a teeny-tiny molecular-sized donut, complete with the hole, or *channel,* through the middle, you'll have a decent visual understanding of how a channel protein looks. Usually the channels are closed, but they can be opened by various ligands or other stimulations. Importantly, **each channel is opened by a different ligand or stimulation.** These proteins enable a cell to control the movement of ions into and out of the cell in response to different inputs. Channel proteins ultimately control the nervous impulse.

In your everyday world, a gate is a door that opens and closes and can block some things from moving through it. We use the term *gating* to indicate the opening and closing of a channel protein. Remember, these proteins are also receptors. They have shapes that allow them to interact with ligands. Those ligands act like keys to either open or close the gate. Ligands are not the only things that can open the gates of channel proteins. Some channel proteins are controlled by physical conditions like voltage, heat, pressure, or light. It's important to remember that these ligands and physical conditions all do the same thing. They interact with the receptor part of the donut, causing the channel to change shape and either open or close the gate. When the gate is opened, ions will move through the gate. This is the heart and soul of nerve cell function. I'll show you several examples of exactly how this works. And since these channels are the essence of nerve function, it should be no surprise that many neurotoxins affect channel proteins.

Neuron depolarization: Doing the "wave"

Okay, so we have a neuron with charges separated across its membrane; it is polarized. The separation of charges has established a voltage across the membrane. The ion channels, or *gates*, are in whatever position they normally are when the cell is at rest. These are all the things the neuron needs to send its message. It's time to put this all together and see how the neuron works its molecular magic.

I'm boiling this down to seven steps, and I'm going to use one of my favorite receptor proteins called the *TRPV1 channel* as an example to lead you through each step in more detail. I'll tell you why this is one of my favorite proteins in the next chapter. For now, all you need to know is that the TRPV1 protein is embedded in the membranes of the dendrites of some neurons, is activated by temperatures above 43°C (109°F), and is part of the reason we can feel heat at all [9].

Step 1 – channel activation: When the TRPV1 channels in your fingertips or other parts of your body are exposed to high temperatures, like maybe you touched a hot element on your stove, the TRPV1 proteins physically **change their shapes so that the ion channel opens.**

Step 2 – depolarization: When the first TRPV1 channel is opened in step 1, positively charged ions (black dots in the illustration below) enter into the neuron and decrease the transmembrane voltage in the nearby vicinity of the first TRPV1 channel activated. It's important to remember that opening the first TRPV1 channel does not depolarize the entire neuron, just the tiny area near the first heat-activated channel.

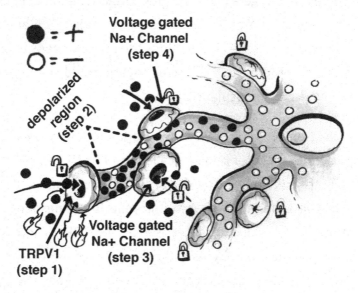

Steps 3-6 – "the wave": Time for a big science term, *voltage-gated sodium channel*. Instead of being temperature-gated like TRPV1, these channels are opened and closed based on transmembrane voltage. What this means is that as long as the voltage is negative inside the cell, these channels remain closed. But, when calcium enters the cell through an opened TRPV1 channel, it neutralizes the negative charge in that first little segment of the neuron. This change in transmembrane voltage is the trigger for the voltage-gated sodium channels to open **(Step 3)**. As the changing voltage causes the very first of these voltage-gated Na⁺ channels to open, more positive ions rush into the cell, further changing the transmembrane voltage **(Step**

4). As these new channels open, the transmembrane voltage changes even more, and additional nearby channels open **(Step 5)**. See where this is going? It's another chain reaction, sort of like we saw in the "falling dominos" of the clotting cascade. In this case, however, the chain reaction is happening along the axon of the elongated neuron. First, only the temperature-gated TRPV1 channel in the dendrites opened, but this quickly triggers voltage-gated Na⁺ channels to open, which trigger more Na⁺ channels, and more Na⁺ channels, and more Na⁺ channels that progressively get nearer and nearer to the axon terminals. It's a wave of channels opening, or a *wave of depolarization.*

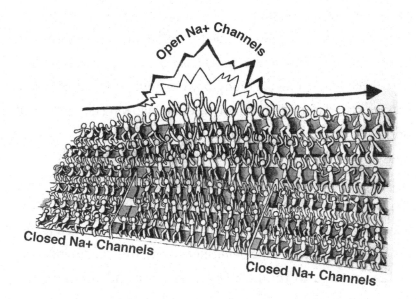

To help imagine what this looks like, think about when you go to a big sports event and people stand up for "the wave" that moves around the stadium. This is a pretty good analogy for how the neuron fires. The first guy to stand up and start the wave is like the first TRPV1 channel opening. The next people to stand up are like the first voltage gated Na⁺ channels opening as the depolarization wave hits them. The Na⁺ channels then progressively open up around the stadium as the depolarization wave hits them in turn. Importantly, the whole neuron (or the whole stadium) does not depolarize at the same time. That would be like everybody standing up in the stadium at the same time... no wave, just chaos. As we will find, there are some toxins that do just that! They cause Na^+ channels in pain-sensing neurons to open up and don't let them close. Sort of like everybody in the stadium standing up and screaming at the top of their lungs after the home team scores. You can imagine the pain inflicted by these toxins!

During depolarization, as positive charges rush in, the transmembrane voltage can, for a brief period, actually flip from being more negative inside the cell to being more positive inside the cell. This is the trigger for **Step 6 – *repolarization of the cell***. After a segment of the neuron is depolarized, voltage-gated potassium channels open, allowing an outflow of positively charged potassium and rapidly repolarizing the cell. These channels are activated by a different voltage than the sodium channels, so they open a few milliseconds later than the sodium channels. When the positively charged potassium leaves this segment, the net positive charge is decreased (same as the net negative charge being increased), the transmembrane voltage returns to the polarized state, and the voltage-gated sodium channels eventually close. In our stadium wave analogy, this is when people sit back down after the wave has passed; the closing of the voltage-gated sodium channels ends the depolarization wave in one segment, but by then, the wave has already moved down to the next segment where the whole process is repeating steps two through six before moving on to the next segment. And on, and on, and on...

Step 7 – neurotransmitter release: Eventually, the depolarization wave reaches the axon terminals. In this final step of neuron firing, neurotransmitters are released to stimulate and open the ion channels in the dendrites of the next neuron and keep the signal moving down the nerve. Neurotransmitters are stored in small *vesicles*, like cellular bubbles, that are piled up inside the cell near the axon terminal region. These vesicles wait for the right trigger to make them release their neurotransmitter into the space between the two neurons, or the *synaptic cleft*. That trigger is the depolarization wave coming their way down from the dendrites. Vesicle release has its own amazing molecular mechanism, perhaps I'll save it for a future book.

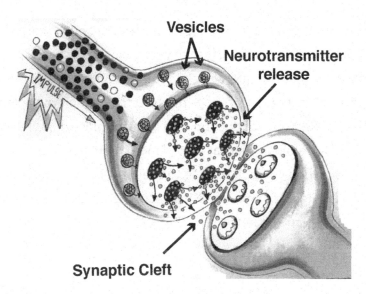

Vesicles

Neurotransmitter release

IMPULSE

Synaptic Cleft

After all this, you hopefully have a conceptual understanding of how a neuron is activated, how the signal travels down the neuron, and how one neuron triggers the next neuron to start the process all over again in seven simplified steps.

Now here is something important that I want to make sure you remember. We are going to be talking about all sorts of toxins and physical conditions that can affect nervous function.

We will be talking later about animal senses and how we hear, see, taste, feel, or smell. **Every one of these things are going to involve steps two through seven**; once it's rolling, the wave of depolarization uses the same channel proteins over and over and doesn't change too much. **What does change is step one, the critical initiation that gets it all started in the first place.** The difference is the receptor protein that is activated to initiate neuronal depolarization. You are going to see a wide variety of initial receptors that give neurons the ability to respond to a wide variety of stimulations. But remember, they are all doing the same thing; they are step one of the process and only serve to start step two and the neuronal wave.

Finally, this same basic process of depolarization can also happen in other cells, especially muscles and hearts, but because of their unique cell shape, nerves are the only cells capable of actually carrying the electric signal **over long distances**. Nervous conduction is an amazing piece of molecular and cellular biology that, at first glance, seems very complex. However, when you boil it down to the most basic components and how these components interact, the "magic" of nerve function becomes science.

What have you learned so far?

Time for a little thinking... I'm a professor and I'm going to do what professors do: I'm going to make you think by giving you a little test to see how well you understand electrical conduction in cells. See if you can come up with some reasonable answers to these thought experiments to make sure you understand the basic mechanism of neuron depolarization.

Question 1: Bee stings hurt. Isn't it amazing how this little creature can cause such pain? How do bees do this? Bee venom contains proteins called *melittin.* When a bee injects its venom, the melittin proteins insert themselves into the membranes of nearby pain-sensing neurons. Melittin is a channel-like protein that doesn't have a gating mechanism. It just pokes a hole into the nerve membrane, and that hole is always open. So, here is the question, **why do bee stings hurt?** Try to combine what I just told you about melittin and what you already learned about neuron function to make a guess.

What did you come up with? If it was something about melittin allowing ions to flow back and forth, thereby causing the neuron to depolarize, you are on the right track. Melittin is an example of what's called a *pore-forming toxin*. These toxins cause a variety of effects by simply poking holes in the cell membrane and letting positively charged ions enter the cell, thus making the cell depolarize. If the neuron is a pain-sensing neuron, the message it sends to the brain is "Ouch! Pain!" Some pore forming toxins are very aggressive and will kill the cells they imbed into. Others like melittin only cause pain. That one wasn't too tough, try this one on for size.

Question 2: One of the key ingredients in the lethal injection used for capital punishment in the USA is nothing more than potassium. True, there are other things in the lethal injection that are designed to first knock a person out and to paralyze their muscles, but the chemical that ultimately causes the heart to stop beating is nothing more than an overdose of potassium in the blood stream. Why does excess potassium stop a person's heart? Here's a hint: we talked about how neurons depolarize, but the basic seven-step mechanism is also used by *pacemaker cells* that control and coordinate the pumping of the chambers in the heart. Think about what you have learned about the depolarization cycle of the neuron and try to come up with a molecular biology-based reason why injecting a massive dose of potassium into somebody's bloodstream can stop electrical conduction in the pacemaker cells of the heart and eventually cause a heart attack.

What did you guess? The answer I came up with has to do with Step 6 of the depolarization process. After sodium rushes into a pacemaker cell during step 5, potassium needs to rush out of the cell to reset the transmembrane voltage and get the cell ready to fire again. During Step 6, potassium **normally** leaves the pacemaker cell because there is **normally** more potassium inside than outside the cell, and potassium diffuses out of the cell following its concentration gradient. But if there is more potassium outside the cell (because of a massive potassium injection), there is no concentration gradient and potassium will not leave. As a result, the pacemaker cell's transmembrane voltage cannot be reset. This sticks, or arrests, the voltage-gated sodium channel gates in the open position and the pacemaker cannot "fire" again. Eventually pacemaker cells stop firing altogether and the heart goes into *asystole,* or cardiac arrest.

Hopefully you did okay on these questions. The ability to extrapolate your knowledge to answer questions to which you don't really know the answer is an important clue that you are starting to understand a topic. For example, you might have heard the phrase "rubbing salt in the wound" or maybe experienced the sting associated with getting salt into a fresh cut. It's common knowledge that getting salt in a wound makes it hurt more, but as far as I can tell, there has never been a published study that actually investigated **why** salt makes wounds hurt so much. Nobody knows for sure! Based on my experience however, my best guess is that salt (NaCl) in a wound causes pain simply by increasing Na^+ ion concentration outside the neuron and causing a more vigorous cellular depolarization. Is my guess correct? I don't know but it is based on what I know about nerve cell depolarization. Hopefully as you learn more, your guesses will become more and more accurate. It's a good indication that you are starting to understand molecular biology.

In the coming chapters we are going to discuss how neurotoxins work, how electric eels generate large electrical voltages, and how your senses of hearing, sight, smell, taste, and touch work. As you start reading about these topics, make sure you remember that all nerves depolarize in essentially the same way. Sure, there are some differences, but for our purposes they are the same. All nerves use steps 2-7 to send signals to the brain but it's that critical step 1 that starts the whole process. Whether a neuron will respond to a stimulation is dependent on whether or not the neuron has a receptor protein that can be activated by a certain stimulation.

Chapter 3: Neurotoxins

Roasted Red Headed Centipedes were a traditional medicine in some cultures. Perhaps because of the pain-killing toxins in the centipede venom.

With a basic understanding of how neurons work, we can resume our discussion of toxins versus medicines and look at some toxins that target the nervous system, or *neurotoxins*. Since the nervous system is linked to so many important functions, toxins that target neurons can have a wide range of effects in people. Some toxins, like those from ants and hornets, cause extreme pain. Other toxins, like those from some centipedes or snails, actually block pain. Still other poisons from some frogs block the neurons involved in breathing. Simply saying that these toxins cause pain, block pain, or stop nerve function, however, really undersells the molecular mechanisms of these interactions; it keeps them in the realm of being magical. My goal in this section is to demystify neurotoxins and help you understand exactly how they work.

Molecular machines

Neurotoxins are ligands that bind to receptors on the surface of your neurons. Frequently these neurotoxin receptors are channel proteins. Neurotoxins most often impact receptor function in one of two ways; either forcing channels to open or forcing channels to close. Opening of the channel causes the neuron to artificially depolarize, while closing the channel prevents the normal ligand from opening the channel when needed. Channels can open and close because just like the machines you see every day that mechanically turn on or off, or open and close, receptor proteins are what I call "molecular machines." They have very simple physical movements and mechanisms that make them mechanically open or close. Instead of being built from metal, they are proteins that are built from amino acids. And, just like any machine, if you throw a monkey wrench into its gears, you are going to goof it up. Try to think of neurotoxins as molecular monkey wrenches that stop these machines from working properly.

I'm going to use the TRPV1 and the $Na_v1.7$ sodium channels as examples since these proteins are some of my favorites and they have some extraordinary and interesting functions in our bodies. But the lessons here could be extrapolated to literally thousands of other proteins in your body or millions of other proteins throughout biology. I'm going to show you actual pictures of these proteins that were captured by X-ray diffraction, the imaging technique I introduced in Chapter 1. These images are snap shots that capture the proteins as they really are (or were) in a moment of time. These images show how these molecular machines actually open and close during neuron firing. Then, finally, we will take a look at how some toxins alter the activity of these channels. If you have ever wondered exactly **how** medicines or toxins work, this next section should give you a really good conceptual understanding.

TRPV1: The human heat sensor protein

The discovery of the TRPV1 (trip-vee-one) channel is a really nice example of what is called *basic science* or *basic research*. Basic science is scientific investigation that does not have an immediate application to human health. There is no obvious or immediate payoff for basic science investigations, so most of the time basic science is funded by grants from various government organizations. And yet, these basic science discoveries often form the foundation upon which future medicines or real-world applications are based. You can think of basic science as the base upon which a pyramid of science is built. At the apex of the pyramid are lifesaving medicines (*applied science*), but they would never have been developed without the basic science that built their foundation.

The scientists who discovered TRPV1 didn't actually set out with the goal of discovering how we feel heat. In fact, they were studying a different topic completely. They were actually interested in knowing why eating hot peppers made peoples' mouths burn. They knew about *capsaicin,* the molecule/ligand in peppers that makes them spicy, and they hypothesized that capsaicin must be binding to a receptor to cause the heat associated with hot peppers. They searched for a channel protein that capsaicin could bind to and activate, and they found TRPV1 [9]. Originally this channel was called the *capsaicin receptor,* but it was later renamed to TRPV1 since its structure was similar to a preexisting family of other TRP channels. After scientists discovered TRPV1 and started studying it in more detail, they also learned about its more important roles in pain and temperature sensation. With this information in hand, more health-orientated research is now underway to design new ligands to control the TRPV1 channel. This is a nice example of basic science that started out just as curiosity but ended up as an important foundation for medically related investigation. There is a highly recommended video of David Julius on the "iBiology" website where he describes the discovery of TRPV1. If you search for "iBiology David Julius" you should find it.

In order to understand how TRPV1 works, we need to zoom in and look at the details of TRPV1 structure. The TRPV1 channel consists of four individual units that are arranged around a central hole. Like I said before, imagine a donut. The TRPV1 channel looks like a donut that was made by a very poor baker. The hole, or channel, is the "business end" of the protein where calcium ions travel through the protein when it is opened by heat or other stimulations. In figure 1 are four images that each show the TRPV1 channel from a different perspective. In the upper left side view, the regions of the protein that sit either within the cell membrane (*transmembrane region*), or actually inside the cell (*cytoplasmic region*), are indicated, but it is difficult to see all four subunits from the side. When viewed from the top, however, (upper right), the four subunits are easily seen (colored in different shades of gray). Notice how these four individual units are arranged around the central pore/channel through which calcium is allowed to pass. See? It's a really badly shaped donut.

Although there are no images of the open versus closed states of TRPV1, the pore would most likely appear slightly bigger in the open state than in the closed state. This allows calcium ions to pass **only** when the protein is activated, and the gates are open. Think of the gate as sort of like a heat-sensitive teeter-totter that is barely weighted towards the "closed" side. The addition of a little heat shifts the balance to the "open" side. It's not precisely known how the heat physically causes the TRPV1 channel to open. My guess would be that the increased temperature simply destabilizes a part of the protein and allows the channel to rearrange itself a little into a conformation that is more stable at the higher temperature. Regardless of my guess being right or wrong, many ion channels work in similar ways to this TRPV1 channel, by simply providing an adjustable channel through which ions can flow into a cell.

In the lower left panel, I have again shown the TRPV1 channel from the side, except this time I took away two of the subunits. This view allows us to see the actual channel as it passes through the protein from top to bottom. You can see how the TRPV1 protein really provides a path for ions from outside to inside the cell. These ions would never be able to get into the cell without this channel since the cell membrane is extremely resistant to letting ions pass through.

In the last figure on the bottom right, I have revealed all the secrets of the TRPV1 channel! I changed how the protein is represented from a *space filling model*, to a *ribbon model* so that you can better see the individual parts of the protein (the amino acids) that are responsible for controlling the flow of ions through TRPV1. Again, we call the control of ions *gating,* and the parts of the protein that control gating are the *gates.* TRPV1 has two such gates, the upper and lower gates [10]. You can see that the gates are nothing more than individual bits of the protein (amino acids) sticking into the channel. These amino acids can move into or out of the channel to open or close the channel. Such gates are common among channel proteins. The inset in the bottom right panel simply shows a blow-up of the TRPV1 channel and its gates.

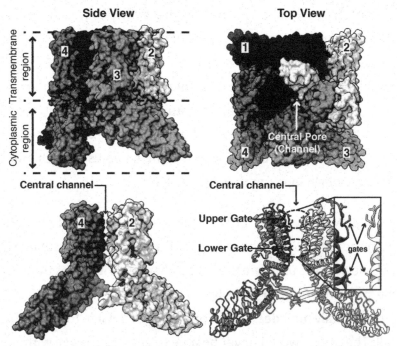

Figure 1: The TRPV1 channel from four perspectives. The upper panels show TRPV1 from the side and top views. The lower left panel has two subunits taken away to reveal the inner channel. The lower right panel is a ribbon model that points out the precise amino acids in the TRPV1 channel that act as gates to control ion flow through the protein.

Chemical heat: Toxins that affect the TRPV1 channel

Here's another thought experiment for you. What do you think would happen if the TRPV1 channel gates were widened? Also, what do you think would happen if the gates were widened and couldn't be closed? How would throwing the gates wide open change the way calcium flows through the channel?

More calcium would get into the cell, right? Well, this is exactly what happens when some ligands that activate the TRPV1 channel bind to it! In figure 2, I have shown another ribbon model of the TRPV1 channel, this time looking down through the channel.

Panel A shows the complete TRPV1 channel, and barely visible inside the central pore are the amino acids that form the lower gate. Panel B is zoomed in on the pore and shows two slightly different versions of TRPV1 that have been superimposed on top of each other. The black ribbons are the normal TRPV1 protein, and the gray ribbons are the TRPV1 protein that was bound to capsaicin. In panel B, you can see how the gray gate amino acids are moved back, enlarging the pore, when capsaicin is bound to the protein. Finally, in panel C, I have taken away all the other parts of the TRPV1 channel except the gates so you can really see how capsaicin changed the position of the gating amino acids.

Remember, capsaicin is the chemical in hot peppers that makes them seem "hot" when we eat them. There is another molecule called *resiniferatoxin* that we also perceive as "superhot". Resiniferatoxin was isolated from a cactus-like plant called *Euphorbia resinifera,* or the resin spurge, and it also produces an intense burning sensation if you are foolish enough to eat it. As a side note, this plant grows in the Atlas Mountains of northern Africa and was used, in ancient times, as a medicine. It is yet another example where knowledge of molecular shapes and molecular interactions can help us to explain what previously seemed to be "black magic".

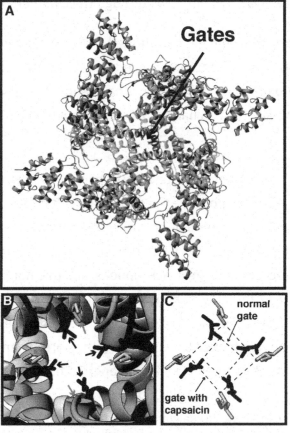

Figure 2: Effect of capsaicin on TRPV1 channel gates. *In panel A, I have shown the TRPV1 channel from the top down. In the very center you can see the central pore and just barely make out some of the amino acids that form part of the channel gates. In panel B, I am showing you two TRPV1 structures that are overlapping each other. The black ribbon is the normal TRPV1 channel plus its gates and the grey ribbon is the TRPV1 channel when capsaicin is bound to it. In panel C, everything except the gating amino acids are removed so you can really see how capsaicin enlarges the gate.*

We have all experienced the painful mouth burn from eating a surprisingly "hot" pepper. So the really interesting question is, **why** are these chemicals so dang "hot" and what does this all have to do with TRPV1? The TRPV1 channel is normally used to sense physical heat, like a hot potato in your mouth, or fire near your skin. With that in mind, it makes total sense that we perceive capsaicin and resiniferatoxin as "hot" even when these molecules are really not physically hot at all! They are just **chemically hot**. But this distinction doesn't matter to your neurons. Your body is fooled into thinking peppers are hot; the nerves can't tell the difference between calcium ions that get into the cell because of real physical heat opening TRPV1 channels and calcium ions that get into the cell because of chemicals, like capsaicin or resiniferatoxin, that artificially open TRPV1 channels. It's all the same to the neurons.

The pain is all in your head, right? Well, yes and no. Like I said, your brain can't tell the difference between a neuron on fire and a neuron being artificially stimulated by a ligand. Capsaicin and resiniferatoxin can't do you any actual damage. Just ask the people who compete in hot pepper eating contests. Nonetheless, it hurts! That initial signal gets transmitted into the exact same part of your brain and your brain processes the signal exactly the same way. We call a molecule that activates, or stimulates, a receptor an *agonist*. Both capsaicin and resiniferatoxin are TRPV1 agonists.

Take one last look at how capsaicin and resiniferatoxin affect the upper and lower gates of TRPV1. In figure 3 on the next page, I am showing you the TRPV1 channel from the side and have measured the actual distance between the amino acids that form the upper and lower TRPV1 gates. You can see that compared to the normal open conformation of TRPV1, resiniferatoxin opens **both** the upper and lower gates while capsaicin only really opens the lower gate. As it turns out, we sense resiniferatoxin as about ten thousand times "hotter" than capsaicin, and the molecular reason is right there in the figure. Opening both gates simply lets in even more calcium ions compared to capsaicin, which results in a stronger, faster depolarization in the neuron. Mystery solved. So the next time you have some spicy chili, or some phantom pepper fries, think about what's going on in your mouth. Think about all those TRPV1 channels popping wide open and causing your nerves to depolarize, and depolarize, and depolarize... Pass the milk please!

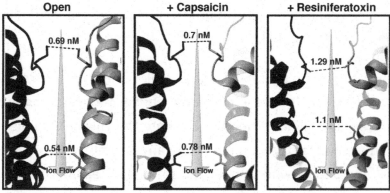

Figure 3: Comparison of capsaicin and resiniferatoxin on TRPV1 gates. Capsaicin and Resiniferatoxin artificially open the TRPV1 gates, fooling the body into thinking there is something very hot in your mouth even when there isn't. Resiniferatoxin is 10,000 times hotter than capsaicin because it opens the top and bottom gates even wider than capsaicin.

TRPV1 toxins: Taking advantage of the system

Since the TRPV1 protein is so important for heat sensation, and since evolution is such an opportunistic force of nature, it isn't surprising that there are many animal toxins that work by targeting the TRPV1 channel to cause pain in their victims. Examples include *double-knot toxin,* produced by the tarantula-like Chinese bird spider [11], *RhTx toxin,* produced by the red-headed centipede [12], *BmPO1 toxin,* produced by the Manchurian scorpion, and a protein called *F13,* produced by the saw-scaled viper [13]. As you might expect, all of these toxins cause an intense burning sensation when they are injected or applied to animals. These toxin proteins probably won't kill any more than capsaicin will, but they will certainly cause pain and scare away aggressors. While it is not yet known exactly how these toxins affect the TRPV1 channel, a good guess would be that they probably either increase the sizes of the upper and/or lower TRPV1 gates, or perhaps simply prevent the gates from closing.

Manchurian
Scorpion

Interestingly, there are other naturally occurring toxins that, instead of activating TRPV1, actually block its function. Molecules that block another protein from functioning are called *antagonist* molecules, and these molecules can be very useful as medicines. An example of a TRPV1 antagonist is the *APHC1-3 toxin* produced by sea anemones [14]. Here again, it is not precisely known how this toxin blocks TRPV1 function, but a good guess would be that it somehow locks the TRPV1 channel into the closed position. Research and development on this toxin is currently ongoing with the hope of producing new medicines to suppress pain. Perhaps another use might be to block TRPV1 channels in your mouth and cheat in your local hot pepper eating contest?

I have one more interesting thing to share about the TRPV1 channel before we move on. Some animals such as birds can eat hot peppers all day with no ill effects. They still have the TRPV1 receptor and can still sense heat, they just aren't affected by capsaicin. Recently, it was discovered that the chicken TRPV1 channel has a mutation in it that makes it insensitive to capsaicin [15]. This mutation falls within the gate region of the protein and probably somehow prevents capsaicin from inducing the change in protein shape that causes the lower gate to open in the normal protein. It's just another nice example of how molecular shapes dictate function.

Okay, so now we have explored the basic molecular biology behind the TRPV1 channel and learned how this protein can initiate signals in the nervous system. The TRPV1 channel is just one of about thirty TRP channels made by humans. Another interesting TRP channel called *TRPM8*, or the *menthol receptor*, is responsible for cold sensation and is activated by menthol and peppermint [16]. It's sort of the opposite of how capsaicin feels hot because it activates TRPV1. Menthol and peppermint feel "cold" because they activate TRPM8. Your brain can make sense of all this because some neurons in your mouth make TRPV1 and others make TRPM8. If TRPV1 neurons are depolarized, they send a message that the brain knows means "hot". If TRPM8 neurons are depolarized, they send a message that the brain knows means "cold". If both TRPV1 and TRPM8 were in the same neuron, the brain wouldn't be able to tell if it was getting a hot or a cold signal. We will encounter another TRP channel later that is important for sensing noxious things, like horseradish and wood smoke [17].

For now, let's turn our attention to another important protein in the nervous system, the $Na_v1.7$ sodium channel, which is required for pain perception in humans.

$Na_v1.7$: The human pain protein

Humans make about nine different types of voltage-gated sodium channel proteins, each with a slightly different function. The $Na_v1.7$ channel is only made in nociceptor neurons that are responsible for sensing pain. During the depolarization of these neurons, the $Na_v1.7$ channel is responsible for carrying the wave of depolarization (steps 3-6) all the way down the neuron. Because of this important function, the $Na_v1.7$ channel is essential for our ability to feel pain and anything that increases or decreases $Na_v1.7$ function will either cause or block pain.

The $Na_v1.7$ channel was originally identified based on a *gain-of-function* mutation in the protein that caused chronic pain in the unlucky person that had the mutation [18]. A gain-of-function mutation is a mutation that causes a protein to activate, even when it shouldn't. If you are a researcher interested in learning about the sensation of pain, it's a really smart move to study people with chronic pain; if you can figure out what **causes** the pain, you will probably learn something about **how we perceive** pain and maybe help these people at the same time. Researchers analyzing DNA from patients with chronic pain syndrome found mutations within an unknown gene. With subsequent analyses, they determined that this piece of DNA codes for the Na$_v$1.7 channel. The mutations in this piece of DNA result in a change in the structure of the protein, which causes the protein to stay open longer than it should, giving rise to prolonged depolarization waves and the patients' experiences of pain in the absence of injury. Basically, this gain-of-function mutation causes a short circuit in neuron function such that neurons depolarize even without an actual pain stimulation.

Interestingly, there are also *loss-of-function mutations* in the Na$_v$1.7 sodium channel that have been documented in the human population. For instance, several members of a family in Pakistan who had no pain sensation at all were found to have loss-of-function mutations in the Na$_v$1.7 protein [19]. One of these individuals was apparently a street performer who could walk on coals and run knives through his arms. This person simply didn't feel pain. This may sound like a quasi-superpower, but pain is actually very important for healthy individuals. Think about how dangerous it would be if you couldn't feel damage to your body! Individuals with loss-of-function mutations in the Na$_v$1.7 sodium channel need to be extra cautious and visually inspect their bodies daily to make sure there is no hidden damage.

Similar to the TRPV1 channel we talked about earlier, the Na$_v$1.7 channel also has a central channel through its middle that allows ions to flow through it and into the cell. This is a common feature on channel proteins and really defines these proteins as a group. As before, this channel is normally closed, but it opens in response to changes in voltage in the cell. Let's pause for a second and think about this; the Na$_v$1.7 channel opens and closes in response to changes in the transmembrane voltage. This is really cool, but how does a protein **sense and respond** to changing cellular voltage in the first place? How does the $Na_v1.7$ molecular machine **physically** work? The answer, as in much of molecular biology, is both simple and elegant.

$Na_v1.7$ has lots of positively charged amino acids around the opening of its channel. Recall your basic physics: opposite charges attract, but like charges repel each other. When a cell depolarizes, the intracellular positive charges increase and **repel the positively charged amino acids** located on the $Na_v1.7$ channel. Amino acids are physically repelled and pushed around, causing the $Na_v1.7$ channel to change shape. It's simple, elegant, and another nice window into how molecular machines like $Na_v1.7$ work. It's just basic physical principles at the molecular level. Since the discovery of the $Na_v1.7$ channel, we have learned quite a bit about how humans feel pain. As you might expect, the $Na_v1.7$ channel has been the subject of lots of research. Scientists are interested in making painkillers that work by blocking its function. By now, you are probably not surprised to also learn that there are lots of toxins that specifically interact with this protein. Let's now focus on toxins that either increase or decrease $Na_v1.7$ function.

$Na_v1.7$ toxins: Stings that open and close the gates

We'll start with the most painful sting on record, according to the *Schmidt sting pain index*. The Schmidt sting pain index is a ranking of the pain associated with various insect stings and bites. Basically, Dr. Justin Schmidt let himself be stung or bitten by many different insects, then ranked the pain from each sting or bite relative to the others [20]. The most painful insect bite, according to Dr. Schmidt, is caused by the bullet ant. Dr. Schmidt describes the bullet ant sting as, "Pure, intense, brilliant pain ... Like walking over flaming charcoal with a three-inch nail embedded in your heel." Ouch! If you want a more visual perspective on the pain caused by a bullet ant, you can also watch Coyote Jones on YouTube as he lets a bullet ant bite him...it looks really painful, no thanks!

The secret to the bullet ant's painful bite is the toxin it injects. The toxin is called *poneratoxin* [21]. This toxin has the interesting ability to bind to $Na_v1.7$ channels and lock them in the open position [22]. Remember that, normally, after sodium channels open (Step 3), the potassium channels open (Step 6), and this resets the neuron by again changing the transmembrane voltage and eventually closing the sodium channels. With poneratoxin bound, the exit of potassium and subsequent decrease in transmembrane voltage does not cause the $Na_v1.7$ channels to close as they normally would. Instead, they stay open and the nerve continues to fire. Go back to the stadium "wave" analogy, but instead of standing up row-by-row, imagine all the sports fans standing up and yelling at the top of their lungs at the same time, and not stopping for hours. The effect would be that, even with very little actual tissue damage, the nerves would be fooled into thinking there was extreme damage. The pain must be extraordinary.

Since most mammals make $Na_v1.7$ channels that are very similar, evolutionary forces probably shaped poneratoxin to activate $Na_v1.7$ to keep invaders away from the bullet ant nest by causing insufferable pain. There are several other toxins that, instead of activating the $Na_v1.7$ channel, cause it to stay closed. In these cases, evolutionary forces probably shaped these toxins such that prey animals could be more easily captured and eaten without panicking the animal.

A good example of this is the previously-mentioned Chinese red-headed centipede. These centipedes can grow to about eight inches, and they eat various bugs and small rodents. Red-headed centipedes have reportedly been used to help cure wounds for centuries in Asia, and there are stories about roasted and ground up centipedes being used for pain relief. Its venom contains many toxins, including the TRPV1 activating toxin, *RhTx*, and, very interestingly, the *Ssm6a toxin*. Ssm6a interacts with the $Na_v1.7$ channel [23]. but instead of activating the channel, Ssm6a causes the channel to get stuck in the closed position. The result is that the channel cannot open, and pain sensation is completely blocked. Sort of like what would happen to a person with a loss-of-function mutation in $Na_v1.7$. When compared to the powerful painkiller morphine, Ssm6a was about 150 times better at blocking some kinds of pain. Ssm6a is also highly specific to the $Na_v1.7$ channel, meaning that low doses of Ssm6a might prove to be a potent and selective pain-inhibiting drug. If toxins like Ssm6a could be developed into medicines, they would have enormous advantages over other opioid painkillers since their high specificity for $Na_v1.7$ would probably make them non-addicting.

By the way, if you are wondering why the red-headed centipede venom contains both the $Na_v1.7$ antagonist Ssm6a, which blocks pain, and the TRPV1 agonist RhTx, which causes pain, so am I. Venom from these centipedes has been found to have about 500 different types of toxins in it, but it is not understood how the various toxin combinations work together to subdue the centipedes' prey. Hopefully this mystery is revealed in the future.

One final toxin we will talk about that also directly modifies activity of the $Na_v1.7$ channel is the *ProTx-II toxin* from the Peruvian green velvet tarantula. Like Ssm6a toxin from the red-headed centipede, the ProTx-II toxin also blocks the flow of sodium through the $Na_v1.7$ channel. There are some excellent X-ray diffraction images that illustrate, again, how toxins can influence channels through increasing or decreasing gate opening and closing [24]. Shown on the next page is the complete $Na_v1.7$ channel. In the top image, you can see that like the TRPV1 channel, the $Na_v1.7$ channel also consists of four subunits that are arranged around a central channel. The bottom two images show the $Na_v1.7$ protein in the open and ProTx-II-blocked states. In the bottom left picture, showing the open channel, notice how, directly in the center, there is a nice clear hole? This is the gate for this channel and it physically narrows or widens to either close or open the channel. This gate is blown up in the lower corner for more clarity. Next, look at the bottom right set of images. These pictures show what happens to the gate when the ProTx-II toxin is added to the $Na_v1.7$ protein. Notice that the gate is deformed and narrow when ProTx-II binds. This closing physically blocks sodium ions from entering the cell, and effectively stops nerve cell function. Drug companies are obviously very interested in this toxin since, if developed properly, ProTx-II might make a good pain killer.

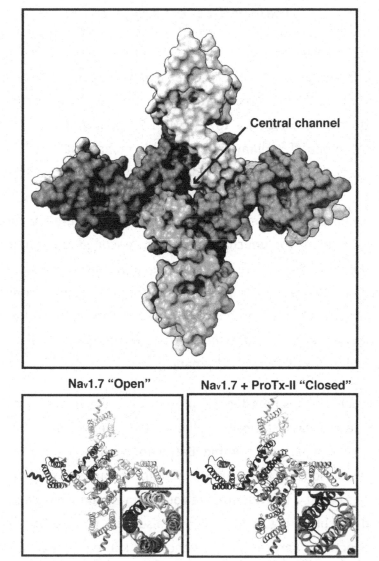

Figure 4: The Na$_v$1.7 protein in the open and ProTx-II blocked states. *The top image shows a top view of the Na$_v$1.7 protein. Like TRPV1 it is built from four identical proteins to make one larger protein with a central channel running through the middle. The bottom left shows Na$_v$1.7 in a ribbon diagram so you can really see the gate that controls ion flow through the central channel. The bottom right shows how the ProTx-II toxin changes the shape of the Na$_v$1.7 gate to block ion flow and stop the protein from functioning.*

The importance of being specific

Part of what makes the ProTx-II toxin so exciting is its high *specificity* for the $Na_v1.7$ channel. ProTx-II is one of the most potent $Na_v1.7$ blockers found so far, and it binds with eighty-fold greater selectivity to the $Na_v1.7$ channel compared to other types of sodium channels. This means that ProTx-II is unlikely to affect other sodium channels in humans. This is important! General inhibition of other sodium channels would likely have dire consequences. Remember that there are about nine different sodium channel proteins encoded in the human genome. Each of these channels has important functions. If a toxin inadvertently disrupts the functions of these other sodium channels, it has the potential to do great harm. This is exactly what happens with *batrachotoxin* and *tetrodotoxin,* two of the deadliest toxins known to humans.

Batrachotoxin is found in the poison dart frogs of the Amazon jungle and was historically used by Amazonian natives to poison arrows and darts for hunting. Batrachotoxin is so deadly that a single poison dart frog contains enough poison to kill ten human adults in just a few minutes. Tetrodotoxin is found in puffer fish and the blue ring octopus and is also very deadly. One blue ring octopus carries enough venom to kill maybe two dozen human adults. Some puffer fish are eaten as sushi and must be prepared by highly trained chefs to avoid accidently releasing the toxin and poisoning customers, which is a poor business model. Like ProTx-II, Ssm6a, and poneratoxin, batrachotoxin and tetrodotoxin also bind to sodium channels including the $Na_v1.7$ channel, but instead of being highly specific for the $Na_v1.7$ channel, these toxins bind to several different kinds of sodium channels. All these channels are similar in structure, and batrachotoxin and tetrodotoxin have the right shapes that allow them to bind to regions of several different sodium channels that are *conserved*, or the same throughout the other sodium channels. Since these other sodium channels are important for general nerve function, including breathing and muscle control, you can probably understand why batrachotoxin and tetrodotoxin are so deadly. These toxins are unlikely to be developed into pain killers any time soon since they would be "the ultimate" painkillers, meaning they would probably kill people.

Batrachotoxin works on sodium channels sort of like capsaicin and resiniferatoxin work on TPRV1, or like the bullet ant's poneratoxin works on the $Na_v1.7$ channel; batrachotoxin locks

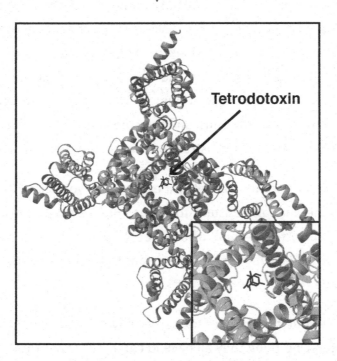

Figure 5: Tetrodotoxin binding into the $Na_v1.7$ channel. Much like a cork in a wine bottle, tetrodotoxin simply blocks the flow of sodium ions through the central channel of the protein.

sodium channels into the open position. Once this happens, the muscles, including the heart, basically lock up and cannot relax. Tetrodotoxin works in the opposite way. Tetrodotoxin prevents sodium channels from opening and causes nerves to stop responding to stimuli. In figure 5 above, you can see a picture of Tetrodotoxin binding into the $Na_v1.7$ channel [25]. Tetrodotoxin acts like a cork in a wine bottle to block sodium ions from flowing through the channel, but in this case, the bottles are many different types of sodium channels. The result is a paralysis of muscles that are involved in moving and breathing. If untreated, both batrachotoxin and tetrodotoxin are deadly.

Toxin-resistant animals

Have you ever watched a video of a mongoose or a honey badger fighting a cobra and winning? Even though the mongoose was bitten during the fight? Think of Rikki-Tikki-Tavi fighting the cobras Nag and Nagaina in the classic Rudyard Kipling short story and, against all odds, winning. These are good examples of animal resistance to toxins, and they present one last chance to use toxins as a tool to illustrate principles of molecular biology.

Here is yet another thought experiment for you! How, or why, are some animals resistant to toxins? Before reading the next paragraphs, try to think of a molecular hypothesis to explain how Rikki-Tikki-Tavi could be resistant to the toxins of Nag and Nagaina.

Think about the various toxins we have talked about in this chapter. One thing they all have in common is that they are all ligands that precisely fit into receptors, and, often, they change the receptor function by changing the receptor shape. Recall the penicillin-resistant penicillin-binding protein in MRSA that we discussed in Chapter 1. With these things in mind, what do you think would happen if you changed the structure of the ligand binding site for a receptor that a toxin normally binds to? Would the toxin still bind to it?

If you said no, you are correct, and this is exactly how some animals are resistant to toxins. They have evolved slightly altered versions of the proteins that are normally attacked by various toxins. A good example of this is the mongoose that is resistant to the cobra toxin *bungarotoxin*. Bungarotoxin is a toxin that binds to a protein called the *acetylcholine receptor* that allows muscle cells to be activated by nervous system transmissions. Blocking the acetylcholine receptor blocks Step 1 of the nerve signaling cascade, causing a loss of muscle control and paralysis; the signal to contract cannot be transmitted! This would certainly be deadly for a mongoose, but these snake hunters have a trick up their sleeve. Researchers have determined that the normal site where bungarotoxin binds to the acetylcholine receptor is slightly altered in the mongoose [26]. This means that, instead of causing paralysis, the bungarotoxin toxin just floats around harmlessly through the mongoose's bloodstream with nothing to bind to. Like penicillin in a human, it is harmless.

Similar changes in toxin binding sites have been described for the honey badger and other toxin resistant mammals as well [27]. These changes are undoubtedly evolutionary adaptations to living in areas with snakes. In the case of the mongoose and the honey badger, these changes put snakes back onto their dinner menu and help these animals live up to their reputations as bad asses! There are lots of great videos online of honey badgers and mongooses fighting cobras.

Another interesting mammal, the opossum, has been known to be resistant to rattlesnake bites, but, until recently, it was unknown how this worked. Instead of a changed toxin-binding site, opossums have evolved several special proteins in their blood that bind to and inactivate rattlesnake toxins, giving the opossum the ability to be bitten by rattlesnakes several times with no ill effects. This is equivalent to the opossum having a "decoy" receptor for the snake toxin in its blood that "sponges up" the toxin before it has a chance to interact with the real receptor. Several of these anti-toxin proteins have been identified, and recently a team of researchers used this information to make a new protein that mimics the possums anti-toxin protein; it is able to bind to rattlesnake venom and block it from interacting with its normal receptor. Injection of this little protein into mice actually protected them from subsequent injections of rattlesnake venom [28]. In places like North America where rattlesnakes are common, this opossum protein could potentially be developed into a new anti-venom medicine to treat people who have been bitten by rattlesnakes. The current anti-venoms for rattlesnake bites are antibodies that are produced in sheep, purified from their blood, and then injected into humans. This new version is produced in bacterial cells, so it would be less expensive to make and would not require animals for its manufacturing.

One last question for this chapter; have you ever wondered why poisonous and toxic animals, like snakes and poison frogs, don't die from their own toxins? Go ahead, make a guess.

I hope you guessed something along the lines of "their receptors are evolved not to bind their toxins." If you did, good guess!

That's basically how it works in the species that have been studied so far.

I've asked you several questions in this chapter to test your knowledge. I hope you did well, or at least better by the end than you were doing in the beginning? I just wanted you to make some guesses. I call this sort of guess *SWAG*, short for Scientifically Wild Assed Guess. Getting good at SWAG means you are starting to get a good general understanding of a topic. Good SWAG means you are at the point where you can begin to make a good guess based on your background, and sometimes that guess might turn out to be correct. If your SWAG was correct this time, congratulations! You are on your way to understanding the world from the perspective of molecular interactions.

We could go on and on looking at pictures of receptor proteins and talking about how various toxins change their receptor's function. Indeed, huge amounts of time are dedicated to understanding, in the finest detail, how various ligands bind to and affect their receptors. When designing new drugs it is crucial to understand how this works because, as we have seen, even one amino acid in the wrong place can have profound effects on a protein's function. For the more casual science lover, what really matters is that you have a good grasp of how ligand-receptor interactions happen and that you can formulate reasonable SWAG. Just keep in mind that, in almost all cases, toxins bind to their receptors because the shapes fit together like the jigsaw puzzle. Ligand binding changes the shape of the receptor, and therefore the way the receptor works. Receptors, including ion channels, are molecular machines with activities that can be changed in ways that make total sense if we can begin to think about how things work at the molecular level. These things are not magic, they're just normal, everyday events at a microscopic level.

If now that we've gone through this whole chapter, I again asked you the question, what is the difference between a toxin and a medicine? What would your SWAG be? Of course, the fact remains that medicines heal and toxins harm. Maybe now you'd add, though, that toxins in small doses can act as medicines, or that medicines in large doses can act as toxins. Hopefully you'd say something about toxins and medicines both being simple molecules that act as ligands for receptors and change their receptor's activity. Regardless of your SWAG, I hope that you are thinking about why and how this works and that you have adopted a molecular perspective of toxins and medicines.

Chapter 4: Bioelectricity

Electric eels use bioelectricity to find and stun their dinner, but also to communicate in the dark muddy waters where they live.

In my opinion, one of the most amazing "superpowers" in the animal world is the ability of electric eels to generate voltages strong enough to stun small fish, or even large animals like horses. But, *how* is the eel able to perform this miracle of biology? In this chapter, my goal is to build upon what you have learned so far about ligands, receptors, neurons, and channel proteins to show you how this really complex and fascinating animal superpower works. The secret lies in the cell and molecular biology of specialized cells in the eel. I said it before, and I'll say it again. It's not magic, it's just biology.

What is electricity?

When people think of electricity, they often think of things like light bulbs and batteries, but they don't usually associate electricity with animals. Despite this, electricity is actually quite common in biology. There are many examples of animals (including humans) that use electricity as a normal part of everyday activities. In order to understand how electricity works in biology, we should make sure we understand some of its basic principles.

In its simplest form, *electricity*, or an *electric current*, is merely the **movement of charged particles**. If you think about this backwards, whenever there are charged particles moving around, you have electricity. This definition of electricity has three key words that we can focus on for now. *Movement, charged,* and *particles.* Let's explain these words in the context of electricity that people are familiar with: electricity moving through a wire.

The *particles* associated with electricity that move through a wire are electrons. Your high school chemistry teacher probably told you that electrons are the parts of the atom that orbit around the atomic nucleus, which contains protons and neutrons. Electrons are the physical means by which negative *charge* is carried in this universe. You see, charge cannot exist on its own, it needs to be attached to a particle, and this is described by the laws and theories of subatomic physics. Negatively charged electrons are always present in the copper atoms that make the wire, but it's not until they *move* that we actually get electricity, or an electric current. But, how do we make electrons move? This is where batteries and generators come in. Batteries and generators supply a *voltage.* The voltage simply provides the force needed to move the electrons from one copper atom to the next copper atom all the way down the wire. This explanation would probably make my physicist friends blush, but for our purposes, it covers the basics. Electricity is simply the movement of charged particles, and voltage is the force that pushes them. As we will see next, electricity in biological systems follows the same definition of electricity; it just achieves it by different, more biological means.

Electricity versus bioelectricity

Electricity in biology is a little different, but it is still electricity, and it adheres to the same definition. We use the word *bioelectricity* to acknowledge the unique nature of electricity in biological systems. So how does bioelectricity differ from electricity? Let's focus on the definition of electricity again: **the movement of charged particles**.

Bioelectricity still requires charged particles. Remember that charge cannot exist on its own, it needs to be associated with a particle. An electric current through a wire is dependent on the negative charges that are attached to electrons. In contrast, most **bioelectric currents are generated by positive charges that are attached to the protons inside sodium or potassium ions.** This is the fundamental difference between electricity and bioelectricity, but it doesn't change the definition! Electricity is the movement of **charged** particles, **not** the movement of **negative** particles.

Okay, so what about movement? How are sodium and potassium ions moved around to produce bioelectricity? Remember what you already know about cellular depolarization. Remember that all cells are polarized (even bacterial cells), meaning that potassium and sodium ions are unequally distributed between the inside and outside of the cell. Normally, there is more potassium inside and more sodium outside. This changes during the depolarization of cells. When a cell depolarizes, voltage-gated sodium channels open and sodium **moves** into the cell following its concentration gradient. Almost immediately afterward, voltage-gated potassium channels open and potassium **moves** out of the cell, also following its concentration gradient.

That's it. Bioelectricity satisfies the basic requirements for electricity when positively charged ions move back and forth across the cell membrane through channel proteins. While normal electricity might travel for hundreds of miles down a wire, the potassium and sodium ions are traveling the very short distance across a cell membrane. But they are still moving.

Hopefully this all sounds familiar to you since this is exactly how neurons send messages to the brain. You might have been told that your nerves are electric, or maybe that nerves use electricity, or something to this effect. This is why! When nerves depolarize, they are creating electricity. That said, the amount of electricity produced by a single neuron depolarizing is very small. So, in order to be useful to the eel, this little electrical discharge needs to be amplified.

Shocking: How eels generate six hundred volts

Let's take this basic idea of cellular depolarization and see how eels use it to produce electric shocks of hundreds of volts. We'll start by comparing a cell to something maybe more familiar, a battery. The figure below shows a very simple diagram comparing how the electrical charges are distributed around a battery, a polarized cell, and a depolarizing cell.

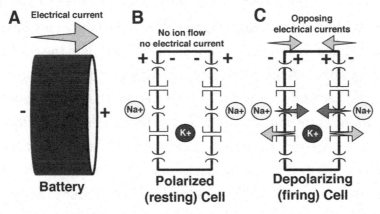

Figure 1: A battery, a polarized cell, and a depolarized cell. (A) A battery has a positive and a negative end, and a voltage across these ends. If you put a wire across the battery ends, you will get an electric current. (B) A polarized cell has more positive charge on the outside and more negative charge on the inside. Since no channels are open, there are no ions moving, and there is no electric current. (C) When a neuron depolarizes, ions flow in opposite directions across the membrane, so the electric currents cancel each other, and the overall voltage is 0.

Notice that the battery has a positive side and a negative side. The separation of charges gives this little button battery ~1.5 volts of energy. It's this separation of charges that allows the battery to produce an electric current if you attach a wire to it; the excess electrons (negative charges) will move through the wire towards the positively charged end of the battery. In comparison, the picture of the polarized cell also has a separation of charge, but it is distributed in a different way.

In the polarized cell, there are more positively charged ions outside the cell compared to inside the cell. This means that the inside is less positive, or, stated another way, is more negative. When the cell is resting, there is no movement of ions and no electric current. When the cell fires, or depolarizes, this distribution of charges is flipped. You can see this in the picture of the depolarizing cell. The sodium is moving into the cell from both sides and the potassium is moving out of the cell on both sides. Each side of the cell experiences an independent electric current as the ions flow back and forth. But, since both sides of the cell are simultaneously allowing sodium and potassium to move, the two opposing electrical currents cancel each other out and no, or very little, electrical current is produced when this cell depolarizes. Think of this as two streams that flow head-to-head towards each other. They will cancel each other out and instead of making a bigger river, they will make a stagnant pool. Nerves depolarize like this. They do produce electric current, but those currents cancel each other out. This is okay for nerves though since the goal is not to make a large electrical current but rather to carry the depolarization wave down the long axon of the nerve.

So, here is the critical question. If the two sides of the cell each make an electric current, but they cancel each other out, how does the electric eel manage to build a strong electric current over the entire length of its body? The answer lies in the positioning of the sodium and potassium channels in the cell membranes of specialized eel cells called *electrocytes.*

Take a good look at figure 2 on the next page. In the first panel I have compared the arrangements of sodium channels and potassium channels in the cell membranes of a normal neuron compared to the electrocytes of electric eels. In the drawing labeled "neuron", the sodium and potassium channels are intermixed on both sides of the cell. This represents the fact, that in most cells, these channels are not localized to any special part of the cell. In contrast, in the electrocyte all the sodium channels are piled up on one side of the cell and all the potassium channels are on the opposite side. This arrangement of channels is critical for making the electrocyte work like a battery!

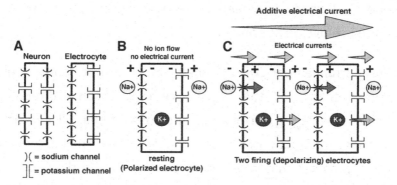

Figure 2: Depolarization of electrocytes. *(A) One of the keys to how electrocytes generate strong voltages is in the arrangement of channels in the cell membrane. Normal cells intermingle their sodium and potassium channels whereas electrocytes put all their sodium channels on one side and all of their potassium channels on the other side. (B) The polarized electrocyte is very much like a normal polarized cell with positive charge on the outside and negative charge on the inside. (C) When an electrocyte depolarizes, all the ions flow in the same direction; they don't cancel each other out. The result is an overall electrical current much like a battery.*

In the middle panel of figure 2 you see an electrocyte that is resting. It is polarized, and it looks very much like the normal polarized cell from figure 1, except for the arrangement of sodium and potassium channels in its cell membranes. Finally, panel C shows an electrocyte caught in the act of depolarizing.

The arrangement of sodium channels along one side means that sodium can **only** enter from the left side. The result is that the left side of the cell briefly becomes more negative on the outside and more positive on the inside, exactly the opposite of what's happening on the right side of the cell. On the right side, where all the potassium channels are located, potassium exits but as it does so, the charge across the cell membrane doesn't change. It remains more negative on the inside and more positive on the outside.

This arrangement of channels means that during the depolarization, all the ions move **in the same direction**. Remember, during the depolarization of a normal cell, ions flow in opposite directions. This might seem like a small difference but remember that when ions flow in opposite directions to each other they make electric currents that **cancel each other out**. In the electrocyte, all the ions are flowing in the same direction, and therefore the currents do not cancel each other out. Instead, they actually add together to make a larger electric current. Think of two rivers that flow together in the same direction. They join their waters and become one larger river. Or think of how you put batteries into a device. The negative side of one battery needs to connect to the positive side of the next battery so the current can flow through both batteries. If you goof up and put two negative or two positive ends together the device won't work because electrical current can't flow. In panel C, there are two electrocytes arranged in the same orientation. When both of these cells depolarize at the exact same time, the voltages of both cells add together. Individually, each electrocyte can make ~0.15 volts (150 millivolts), but when they fire at the exact same time, they add together to make ~0.3 volts, twice as much voltage.

When it comes right down to it, electric eels are built very much like a big stack of AA batteries. Sure, they have fins and heads and organs like any other fish, but what makes these animals truly unique are the electricity-generating organs that run for about two-thirds of the animal's length. There are several different electricity-generating organs, but the primary organ that produces the largest shocks is called the *main organ*. If you could look at the main organ under a microscope, you would see that the individual electrocytes are all arranged in parallel rows.

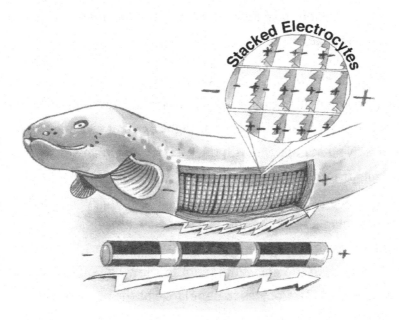

In the schematic representation on the next page, notice the organization of the electrocytes. They are stacked so that the current can continually flow through all of them, just like a stack of batteries. As a matter of fact, when Alessandro Volta made his first battery in 1800 (originally called a *voltaic pile*) he based the design on the main organ of the electric eel!

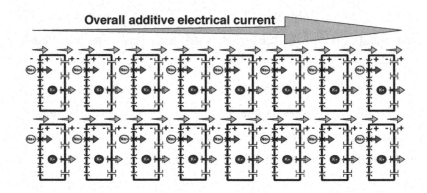

Figure 3: Stacked electrocytes, similar to the arrangement in the main organ. *Thousands of electrocytes stacked up, all in the same orientation, add their small voltages together to create a massive voltage.*

What happens when you put a bunch of batteries in a row and connect them all together? The batteries all add their voltages together to make a stronger flashlight or whatever, right? This is exactly what happens with the electrocytes in an electric eel, except in the eel there are several thousand electrocytes all lined up in multiple rows. That's a lot of electrocytes! And here is something else to consider: all those electrocytes need to be precisely controlled so that they fire at the exact same time, and only when the eel wants them to. How do you coordinate thousands of electrocytes to depolarize at the same time?

An on-off switch for the electric eel

The strongest electric eel discharge ever reported was 860 volts [29]. This voltage can be maintained by the eel for only about a millisecond at a time, but the electric discharge is repeated over and over at a frequency of hundreds of cycles per second. The eel basically does the same thing that a Taser does! This is pretty cool, but it also poses a significant problem. Since each individual discharge is so short, all the electrocytes need to be turned on and off at precisely the same time. If the individual electrocytes are off by even a couple of milliseconds, the overall voltage will be significantly decreased. So how do you make thousands of individual electrocytes discharge at exactly the same moment? Luckily for the eel, this is a perfect task for the nervous system and ligand-gated channels.

To understand how thousands of electrocytes are coordinated, it might help to know that the electric organs of electric eels actually evolved from muscle cells, so they share several characteristics. First, muscles are also composed of thousands of individual cells that need to be simultaneously stimulated to make the whole muscle work as a unit. If your individual muscle cells didn't work as a unit, your muscles would be significantly weaker. Second, the mechanism that coordinates muscle contraction is the same mechanism that controls electrocyte depolarization; every single cell in a muscle (and every single electrocyte in an electric organ) is **individually connected to an input from the nervous system**.

Remember, nerves are made up of neurons, so the signal to "contract" or "fire", which starts in the brain, is transmitted to each individual neuron. Those signals reach the axon terminals of the final neurons of the nerve at the same time. Those neurons simultaneously release a neurotransmitter called *acetylcholine*, essentially bathing the nearby muscle cells or electrocytes in acetylcholine. Acetylcholine is the ligand that binds to the acetylcholine receptor, which we talked about in the last chapter (when we discussed the mongoose that is resistant to the cobra's bungarotoxin.) Acetylcholine works just like any other ligand that causes a channel receptor to open. When the receptor opens, it allows a trickle of positively charged calcium and sodium to enter into the electrocyte, which triggers the same seven-step depolarization cycle we talked about for neurons. The result is that all the electrocytes in the eel discharge simultaneously and create a large shock. To us this sounds amazing, but from the eel's perspective, this is probably no more difficult that flexing a muscle.

Electric shocks: Not just for killing

Electric eels do not always discharge their electric organs at full blast. They can be far subtler than this. Electric eels often live in muddy water with zero visibility and use their electric fields for locating their food and for communication. To find small prey animals hiding among rocks and whatnot, the eels use short, double bursts of electric energy to cause the hiding fish to involuntarily twitch and spasm [30]. This works because the initial electric jolts are strong enough to trigger the opening of the voltage gated sodium and potassium channels in the neurons and muscle cells of the nearby fish, resulting in cellular depolarization and muscle twitching. To the prey this probably feels something like when your doctor "tests your reflexes" by tapping under your kneecap with a rubber hammer, artificially activating the nerve and making your leg kick out. The poor little fish can't resist the eel's searching probe, and it gives away its position by twitching. In the muddy water, though, the eel can't actually see the fish; it relies on its electrical system to detect its twitching dinner.

Seeing in the dark

To detect the presence of the twitching fish, eels rely on

electroreceptor organs located under the eel's skin. These *electroreceptors* are scattered all over the fish's body but are concentrated on its head. While it is known that electric eels can detect electric fields [31], we have not yet discovered exactly how this is accomplished at the molecular level in eels.

In 2017 the mystery of how some animals sense electric fields was at least partly solved in another fish that can also sense electric fields. A voltage-gated channel protein called $Ca_v1.3$ was discovered in the ray-like skate that lives in the ocean. $Ca_v1.3$ was found in special organs called *ampullae of Lorenzini* on the skate's skin and was shown to be responsible for sensing electric fields [32]. The skate has many ampullae of Lorenzini along its body, so multiple, simultaneous messages can probably be interpreted in the brain to inform the skate about the location and/or shape of the electric field that it's detecting. The molecular secret within the $Ca_v1.3$ protein that lets it detect electric fields appears to be the presence of several positively charged regions of the protein that interact with external electric fields. Interactions between these charged regions of the $Ca_v1.3$ channel and surrounding electric fields probably cause a distortion and opening of the protein channel, similar to how voltage-gated channels in neurons sense and respond to changes in transmembrane voltages. The current SWAG is that $Ca_v1.3$ is activated by an electric field, opens up its channel, causes the cell to depolarize, and sends a message back to the fish's brain.

Interestingly, many animals (even humans) make proteins similar to the skate $Ca_v1.3$ channel protein, but these similar proteins lack the positively charged regions found on the skate channel. When researchers used some genetic engineering tricks to put these charged regions onto an otherwise voltage-insensitive channel, that channel gained the ability to respond to electric fields [32]! This experiment helped to prove that the channel relies on those positively charged regions for sensing electric fields.

It has not yet been determined if a comparable protein is also found in electric eels, but in 2018 a similar voltage-gated calcium channel was discovered in a kind of shark called a *catshark* [33]. I won't be surprised at all if a similar channel is discovered in electric eels soon.

Whether or not electric eels have the modified $Ca_v1.3$ protein, we do know that electroreceptors on the eel are activated by its own electric field. Like the skate, these electroreceptors are connected to the eel's nervous system, which probably means that the eel can detect the shape of its own electric field. If there is a small animal inside the electric field it will cause a distortion in the field that the eel can sense, and then find. It is also possible that the electric eel can directly detect the electric field of its prey, although this has not been tested.

Obviously, humans can't detect electric fields so it's difficult to understand what this sixth sense of detecting electric fields might "feel" like, but it's probably similar to any other sense like hearing or vision. Think about how you can use your ears to find something making a sound in a dark room. You can roughly tell what direction the sound is coming from. This is probably the way eels use distortions of their electric field to locate prey they cannot see. While we have only two ears, however, eels have many electroreceptor organs and undoubtedly do better at finding animals to eat in the dark than we would if we relied only on our ears to guide us around a dark room.

Once the eel locates something it wants to eat, it discharges the electric organ in "Taser mode" to stun the animal. But this only stuns the dinner. And sometimes the stun isn't enough to completely immobilize it, so, if the eel doesn't act quickly, the prey has a chance to escape. In this situation, the eel has another trick up its "sleeve"; it can increase the intensity of its electric shock. Eels bend themselves into a C shape, bringing their tails (negative end) and heads (positive end) close together with an unlucky animal in the gap between head and tail [31]. By doing this, the electric field is intensified to at least double the field strength of a non-curled eel. The rapid muscle spasms are thought to exhaust the fish so it cannot struggle and is an easy meal for the eel. You can look up videos of the eels doing this, it's really amazing. Especially once you understand the molecular biology going on!

Think about this experience from the point of view of the poor prey animal. There it is, an innocent minnow hiding away in the weeds, trying to go unnoticed by the marauding eels swimming nearby, when suddenly the electric eel discharges its searching jolt and forces the fish to give away its position with an uncontrollable muscle spasm! A moment later, the minnow is wriggling uncontrollably, until it becomes paralyzed. It is powerless as the electric eel opens its giant maw and swallows it whole (and alive!) Then, fade to black... Doesn't it remind you of a scene from a horror movie, as an exhausted, terrified, heavy-breathing victim tries to hide under the bed? The natural world can certainly be an unforgiving place!

A voice in the dark

One final note about bioelectricity in electric eels is that, in addition to hunting with electrical fields, eels are also thought to use their fields to communicate and identify themselves in the murky depths. Just like humans have distinctive voices that we can use to identify individuals we know within a crowd of strangers. Different species of electric eels might use fields of different strengths and frequencies to identify themselves without being able to visually see one another. In the words of the science writer, Ed Yong, "The electric eel is a battery, Taser, remote control, and tracking device, all in one."

Before moving on, lets recap this discussion of electric eels and bioelectricity. Electricity is defined as the moving of charged particles. Bioelectricity is generated when sodium and potassium ions are moved in or out of the cell, across the cell membrane, through channels. Since sodium and potassium move like this in all cells, all cells are electric to some degree. We saw previously that neurons are specialized for carrying electrical impulses as waves of depolarization all the way down their axons. This is why we say that nervous tissues are electrical tissues. And yet, nervous tissues only generate a few millivolts of electricity, which is nothing compared to the 860 volts produced by some electric eels. One major difference between eels and nerves is the arrangement of electrocytes in the electric organs of eels, each producing a few millivolts but stacking together to produce a massive cumulative voltage. Another crucial difference between electrocytes and neurons is the arrangement of sodium channels on one side of electrocytes and potassium channels on the opposite side. This arrangement produces a voltage across the whole cell and amplifies the electrocytes' ability to make electricity. When it comes right down to it, nerves and electric eel organs are highly similar. They both use similar mechanisms to make electricity; it's just that the cells and channels are assembled differently in the two tissues.

Biological repurposing

Recycling of things like muscles to make electric organs is common in nature. One of the amazing things about biology is that unique solutions to common problems are often based, not on new genes, but rather on the repurposing of "old" genes, proteins, or tissues into new creations. Mother Nature likes to repurpose! In 2014, scientists compared the DNA of different fish that have electric organs and discovered that electric organs in fish have evolved independently at least six different times on Earth [34]. This means that the electric fish do not all have one common recent ancestor. Instead, these fish are all evolved from different families of fish.

This is really interesting, and at first glance it seems to break the basic rules of evolution. One might reason that something so evolutionary beneficial as the electric organ would be highly conserved among fish, and that there would be strong genetic similarities among fish that have electric organs. As it turns out, the electric organ is **so** beneficial to fish, that it has independently evolved many times, each time by converting muscle cells into electrocytes! The electric organ is an example of another kind of evolution, something biologists call *convergent evolution*. Convergent evolution happens when different, unrelated animals that experience and interact with similar environments evolve similar solutions to deal with the common challenges of their environments. Look it up online. You will find many other examples of convergent evolution, and all are fascinating.

Intermission II: Senses and Sensibilities

Many of us take our ability to see, hear, smell, taste, and feel our world for granted. But stop and think about your senses for a second. Exactly how do they actually work? How do your ears convert sound waves into neural impulses? How do your eyes catch light and relay this information to the brain? When you run your finger over a surface, how do you feel it? Many of us have a rudimentary idea of how these things work, we're taught the basics in school. I bet you came up with some version of an image of the ear drum in your ear, the optic nerves connected to your eyes, perhaps some rods and cones. Maybe nerve endings in your fingers? Let's take it a step further. What molecular events actually trigger those nervous impulses? Prepare to be amazed, because the next several chapters are going to focus on the molecular biology that give your ears, eyes, and other sensory organs the ability to sense and interact with their environments. I think you will find that they are both astonishing and elegant.

Sensory system anatomy

Before I move into these next chapters about how your senses work, I want to give you a quick tutorial on the anatomy of a sensory system since they all have some basic characteristics in common.

Your brain only speaks one "language": the language of electrical impulses. When you boil everything down, all of our sensory systems are simply translation systems that receive information about your environment and translate this information into the electrical messages that the brain decodes, interprets, and responds to. These environmental stimulations include physical pressure (touch), sound waves (hearing), light (vision), and the various molecules in the air you breath (smell) and in the food you eat (taste). Here and there we will also discuss some environmental stimulations that humans cannot perceive but some animals can, like electrical fields, magnetic fields, heat, and ultrasonic sounds to name a few. All of these things are simply environmental stimulations that need to be translated into electrical information so that the brain can respond to them. **The secret to whether or not we can perceive a given stimulation is simply the presence or absence of a receptor protein that can interact with that stimulation.**

In all sensory systems, translation of environmental stimuli is achieved with a two-part system that consists of receptor proteins and the nerves that contain those receptors. Receptor proteins can be activated by the physical world, and they are housed within a cell that is, in one way or another, directly connected to the nervous system. Activation of the receptors causes depolarization of the nervous tissue, sending a message in the electrical language of the brain. This receptor+neuron, two-part system is at the core of all our sensory systems. Each sensory system is unique because of its specialized and unique receptor proteins.

Each neuron typically makes only one kind of receptor protein at a time. This is important because all neurons depolarize the same way **no matter what stimulus they might be detecting**. Imagine what would happen if a neuron made two different receptor proteins for example: one that was activated by heat and one that was activated by cold, and that neuron was sending

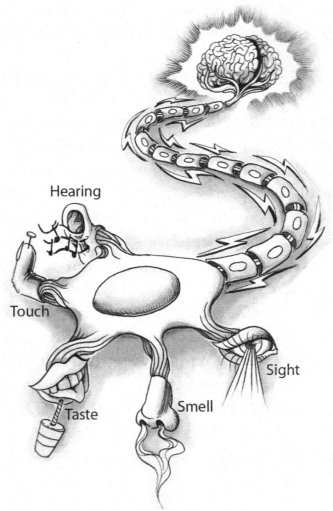

Figure 1: General anatomy of sensory systems. All sensory systems are similar since they all communicate with the brain through the nervous system. Each sensory system is different because each sensory system uses unique receptors to sense and respond to the environment.

a signal to the brain. From the point of view of the brain, all it would know was that a nerve sent it a signal that needed to be interpreted. But the brain would have no way to discern which receptor had been activated and was responsible for the message in the first place. The brain wouldn't know whether it should pull the hand back from a hot stove, or to stay where it was and enjoy the cool water. It would be total confusion.

Instead, since each sensory nerve only makes one type of receptor, the brain knows that when a certain neuron sends it a message, that message means something, like heat, or a pleasant smell, or pressure on a fingertip. Of course, this also means that the brain is constantly receiving unknown numbers of signals that it has to simultaneously decode and respond to. I like to imagine the brain as a little person sitting inside your skull at a switchboard with millions of little switches and lights that are buzzing and beeping constantly on the board. Each little light represents a nervous signal that the poor little brain-person needs to respond to within a fraction of a second. Think about how fast your sensory systems need to work in order for you to see the letters on this page, send the message to the brain, and have the brain construct meanings from the letters. It's actually quite amazing to think of your brain like this. Understanding all the details of how the brain works is one of the biggest mysteries that remain in understanding how animals work.

That's it. Sensory systems are, at their core, extremely simple. Of course, there are plenty of details, and when you really dive deep, they get more complex. In the next few chapters, I'm going to explain how things like sound waves, light, and pressure are translated into electrical signals. Many sensory systems also have specialized cell structures that help to maximize their sensitivity. I hope to help you appreciate how these systems are built so you can be as genuinely amazed as I am when I learn about these things.

Chapter 5: Touch

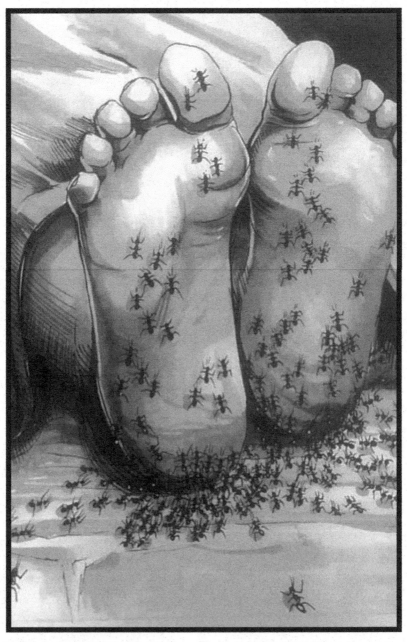

Talk about making your skin crawl! Our sense of touch can be both pleasant and horrible.

Next to vision, I think touch is probably our most important sense. The sense of touch is simply amazing. It lets us tell the difference between sandpaper and an animals soft coat. We can feel the smallest bump on an otherwise smooth surface, and many people can actually read with their fingertips! But, have you ever stopped to wonder how your sense of touch actually works? It's an amazing and elegant system that revolves around a receptor protein that is activated by physical distortion!

You hurt, my feeling!

Two of the anatomical features that make humans a dominant species on Earth are our hands and our brains. Our hands, when combined with our brains, are capable of making almost anything we can imagine. Opposable thumbs and four long, dexterous fingers arguably make our hands second to none on Earth. But imagine if you couldn't feel anything in your fingertips. Imagine how it feels when you wake up in the middle of the night with no feeling in your arms. How useful do you think your hands would be without their delicate sense of touch?

You may have been taught that the sensitivity of a tissue is due to the number of "nerve endings" in that tissue. It is true that different parts of your body have variable numbers of nerve endings, and that the density of nerve endings is directly related to how sensitive that part of your body is. Your hands and fingertips are among the tissues in your body with the most nerve endings, and we absolutely depend on the tactile sensations we get from these nerve endings when we touch or handle different objects.

To understand why the density of nerve endings in a tissue is important, think of each nerve ending as a single pixel in a TV. The more pixels you have in a TV screen, the better the resolution of the picture. Imagine if your fingers each had only one nerve ending. The ability of your fingers to accurately discriminate between slightly different surfaces would be very low. In comparison, if you have thousands of nerve endings, each individual ending can send a signal to the brain where this information can be processed to get a more accurate perception of whatever you are feeling. Like a high-resolution TV or computer monitor. If we mapped our bodies so that the size of our various parts was related to the number of nerve endings in those parts, we would look something like the somewhat disturbing image, called a *sensory homunculus,* shown in figure 1. Obviously, the hands and lips are very large because these tissues have the greatest density of nerve endings. That great density of nerve endings gives us the incredible ability to delicately manipulate the world around us. It also comes with a cost...it's no wonder those little paper cuts hurt so much!

Figure 1: Sensory homunculus. *Different parts of the body are drawn in proportion to the number of nerve endings in the various parts of the body. The hands, lips, and face have many nerve endings, while the legs and feet have comparatively few.*

Under your skin: A quick review of touch sensory structures

So, what really are these so-called nerve endings? What are the structural and molecular mechanisms that actually allow these nerves to respond to touch? As the name implies, nerve endings are structures at the end of sensory nerves in

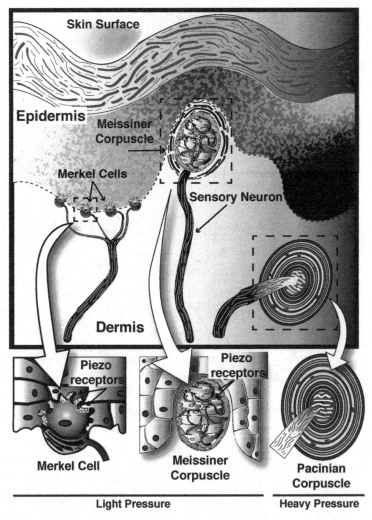

Figure 2: Touch receptors in the skin. Merkel cells and Meissiner's corpuscles are near the surface of the skin and are sensitive to light touch. Pacinian corpuscles are deeper in the skin and are only sensitive to heavy pressure.

your skin. The structures, which are actually bundles of cells, are called *Merkel's Cells*, *Meissner's corpuscles*, or *Pacinian corpuscles*. As shown in figure 2 to the right, Merkel's disks are located close to the surface of your skin, right on the interface between the epidermis and the dermis. They allow you to feel very light touch. Meissner's corpuscles are located just a little deeper, in the top part of the dermis, and are also responsible for sensing light pressure. These corpuscles are comprised of a sensory nerve ending surrounded by an actual capsule made up of other supportive cells and connective tissue. When the capsule experiences pressure, the pressure is transferred to the nerve ending, and receptors in the nerve ending respond. Pacinian corpuscles are found much deeper in the skin and are responsible for detecting deep pressure. Their structure is similar to a Meissner's corpuscle, with a nerve ending surrounded by a sheath of concentric layers of supportive cells. The nerve ending has its own sensory receptors that respond when the nerve ending gets stretched or pushed on by the surrounding concentric layers. We have known that Merkel's disks, Meissner's corpuscles, and Pacinian corpuscles are responsible for our sense of touch for more than 100 years, but it wasn't until recently that we finally uncovered their molecular secrets.

Mechanoreceptors: Meet the Piezos

In 2010, after a long search, scientists discovered two proteins they called *Piezo-1* and *Peizo-2* that are activated by pressure [35]. The word *Piezo* is actually based on a Greek word that means "to squeeze or press." To help understand how these proteins work, figure 3 shows what Piezo-2 actually looks like [36].

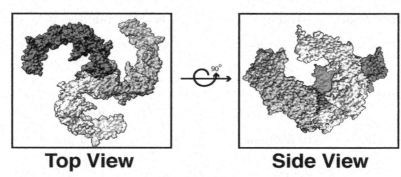

Top View   **Side View**

Figure 3: Structure of the Piezo-2 mechanoreceptor. Viewed from the top, the Piezo-2 protein has three arms that come together around a central channel. The channel cannot be seen in this model. Viewed from the side, the Piezo-2 protein looks like a three-armed tripod turned upside-down. The inverted tripod structure forms dimples in the surface of cells that probably look something like the dimples on a golf ball.

In the figure on the left, you can see that the complete Piezo-2 channel receptor is actually composed of three identical arms that come together at a central point. The figure on the right has been rotated 90° so that you are seeing the Piezo-2 protein from the side. From this angle, you can see that the three arms are sort of projecting upwards from the central point. This particular picture was captured while the channel was in its closed state, so you can't see the pore through the middle in this image. If it were open, the central point where the individual arms come together would have a pore running through it.

Like the other channel proteins that we have looked at, this Piezo-2 central pore is where sodium flows through after Piezo activation. Sound familiar? The cool thing is that, whereas the other channel proteins we looked at were activated by voltage or ligands, **Piezo channel proteins are activated by distortion of their molecular structure**. They are what we call *mechanoreceptors;* they are receptors for mechanical stimulation. Viewed from the side, the individual proteins resemble an inverted domed structure with three arms, where the top of the dome is lower than the edges. In the surface of the cell, they probably look a little like the dimples on a golf ball. This molecular dimple is the secret to how Piezo proteins are activated by pressure. Figure 4 below shows a model of the Piezo-2 protein. From the top, you can again see the three arms, or blades, and a "cap" structure which the pore runs through. The receptor is imbedded into the cell membrane such that the blades are in the same plane as the membrane. The side view image on the bottom right demonstrates the dimple that the inactive Piezo-2 proteins form. Pressure on the cell membrane stretches and distorts the membrane and causes the flexible Piezo protein to bend and flatten out. This bending acts in a

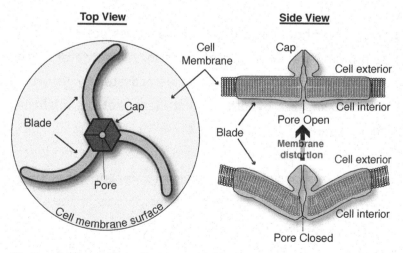

Figure 4: A model of Piezo-2. *The Piezo-2 protein sits in the cell membrane and forms dimples in the cell surface when the receptor is closed. Stretching of the cell membrane causes membrane distortion and forces Piezo-2 to change shape to sit flat on the cell surface. When the protein changes shape, the central pore/channel is opened and ions enter the cell, causing cellular depolarization.*

lever-like way to open up the pore, allowing sodium ions to flow into the cell and change the transmembrane potential [37].

This is exquisite! A simple molecular machine with a shape that is easily distorted by pressure to open it up. Amazingly, a Piezo protein only has to be deformed by about ten nanometers, or one ten-thousandth of a millimeter, to be activated [38]. For orientation, one millimeter is roughly the length of the dash at the end of this sentence - . Imagine a distance that is ten thousand times smaller than that dash. No wonder our fingertips are so sensitive! Through this simple mechanism, the Piezo proteins are able to convert a mechanical force into a cell membrane depolarization, and this is happening every single time you feel anything!

Knock it out: Innocent until proven guilty

Understanding how Piezo proteins work does not prove that these proteins are, in fact, responsible for our sense of touch. At best, knowing that Piezo proteins are activated by distortion allows us to make a hypothesis that Piezo proteins **might** be touch receptors. To prove that these proteins are, in fact, responsible for the sensation of touch, we need more evidence. It's sort of like a lawyer trying to convict somebody of a crime; the lawyer needs to prove that the person was at the crime scene, had a motive, and had blood on his hands. Scientists also need to gather multiple pieces of evidence to "convict" Piezo proteins of being our touch receptors.

To get this evidence, scientists performed an experiment in mice where they *knocked out*, or deleted, the gene in the mouse's DNA that codes for Piezo-2 proteins in the skin, then tested whether or not the mice could still feel touch. Experiments like this are pretty common these days, thanks to a growing number of genetic tools that scientists can use. The scientists compared the Piezo-2 knockout mice to normal mice and showed that the knockout animals were unable to feel touch [39]. In addition, Piezo proteins have been found in the nerve endings of Merkel's disks and Meissner's corpuscles, which prove that they were "at the scene" [40]. Interestingly, Piezo proteins have **not** been found in Pacinian corpuscles, so there may be additional mechanoreceptors waiting to be discovered. Nonetheless, the collected data shows that Piezo proteins are activated by pressure, are required for light touch sensations, and are located in the organs required for touch sensation. Based on the evidence, most juries would probably convict Piezo-2 as being a touch receptor.

Mechanosensors everywhere!

The story of the Piezo receptors doesn't end with touch reception. In fact, this story is just getting started. As scientists dug deeper, they found Piezo-2, not only in Merkel's disks and Meissner's corpuscles, but also in many other tissues, including blood vessels. Scientists had been searching for *baroreceptors*, or pressure receptors, in the circulatory system for years. Baroreceptors are essential for monitoring blood pressure and relaying blood pressure information back to the nervous system. The nervous system uses this information to adjust blood pressure up or down as needed. We've known that baroreceptors exist for years, but the identity of a receptor protein that could actually sense, or interact with, changes in blood pressure kept eluding us.

Finally, in 2018, Piezo-2 was found in blood vessels and was shown to be the long-sought blood pressure receptor protein [41]. It turns out that the baroreceptor is not technically a baroreceptor at all, as it does not directly sense pressure. Rather, it is a mechanoreceptor, which indirectly senses pressure by detecting stretching of blood vessel walls. As blood pressure rises the vessels stretch, causing distortion and activation of Piezo-2. The open channel allows ions to rush into the cell and start a depolarization event, sending a signal to the brain that the pressure is too high. The opposite happens when blood pressure drops; the vessels relax and the Piezo-2 channels return to their closed state, halting the signal. In retrospect, Piezo receptors are the perfect molecules to sense the stretch of cell membranes caused by high blood pressure inside blood vessels and report this information to the body. As before, to prove that Piezo receptors are important for controlling blood pressure, a genetic deletion was performed. The mice without Piezo receptors in their blood vessels were not able to regulate their blood pressure the same as the unaltered animals.

Another fun example of what Piezo receptors do for us is *proprioception*. Basically, proprioception is a sensory mechanism in animals that allows us to know where our bodies are in space even when we cannot see ourselves. It takes time to learn to control our bodies. Think of how clumsy you were in your early teen years. This is because your body was changing faster than your brain could learn to control it. Over time we all get comfortable in our own bodies. We learn how to control our bodies based on the feelings we get from muscles flexing, skin or tendons stretching, or joints bending. All these signals feed into our central proprioception center, and eventually we can automatically interpret the signals and know where our bodies are in space.

Close your eyes and stretch out your arms as far as they will go to each side of your body. Then, with your eyes still closed, try to bring the tips of your index fingers together without touching them. Really, try it! Get your fingers as close as you dare to, but do not let them touch! I bet your fingers got about an inch apart before you were sure they were going to touch. This is proprioception; knowing where your body is in space without needing to see where it is.

As it turns out, Piezo receptors are also found in so-called *proprioceptive neurons* [42]. As before, Piezo receptors sense the bending, stretching, and pressure associated with moving your body and, if they are activated, cause a depolarization event, sending a signal to your brain. The central nervous system receives multitudes of these signals and has to learn to interpret what they mean. Over your lifetime, you have learned to control your body based on probably billions of proprioceptive signals. To demonstrate that Peizo-2 is involved in proprioception, scientists again deleted Peizo-2 from proprioceptive neurons in mice and found that the mice were highly uncoordinated; they all looked as if they had drunk too much alcohol. Without the Piezo-2 receptor, they lost the ability to know where their bodies were. By the way, if you're curious, there is no evidence that alcohol can effect Peizo-2 protein function at this point and it does not appear that the loss of coordination associated with drinking too much alcohol is linked to Piezo-2 function.

Finally, Piezo receptors are also located in your bladder. They seem to be responsible for sensing the fullness of your bladder, not by directly sensing how much pee is in there, but by sensing the degree of stretch in the bladder wall [43]. Going forward, I won't be surprised if we find roles for Piezo receptors in other bodily functions as well!

Before we move on to the next chapter, let's put what you just learned into perspective. So far, we have discussed receptors that give cells the ability to sense stimuli like voltage, temperature, and various ligands. Piezo provides another example of how cells can sense another stimulus they experience every day: physical force. Piezo proteins are mechanoreceptors. At this point, we can rightly extrapolate that pretty much any physical stimulus we can sense is going to have a receptor of some sort that it can activate. In all cases, these receptors just convert the physical stimulus into a nervous impulse that goes back to the brain. The real beauty is in the simplicity of the molecular mechanism. Up next, let's take a look into your ears.

Chapter 6: Hearing

The evolutionary war between bats and bugs is partially waged in the ultrasonic sound range that humans cannot even detect.

If I asked you to guess how your sense of hearing works I bet that, after reading the previous chapters, you could come up with a decent SWAG. By now you should be able to guess that there is going to be a receptor that somehow senses sound, then sends a nervous signal back to the brain. That's all true, but the ear presents some really interesting problems that have been solved in clever ways. For instance, we not only have to be able to hear things, we also need to be able to detect different pitches and different intensities of sound. To solve these more nuanced situations, the ear has evolved some truly sophisticated features.

Do Re Mi Fa Sol La Ti Do!

Sound waves are a complex kind of stimulation that are more diverse than things like pressure, temperature, or a simple ligand. They come in a wide range of both *pitch* (highness or lowness) and *intensity* (loudness). Basically, really fast vibrations make sounds with high pitches and slower vibrations make sounds with lower pitches. The number of vibrations per second is the frequency of the sound. One vibration per second is equal to one hertz (Hz). Humans can hear sounds with frequencies in the range of something like 20 Hz (twenty sound waves per second) to 20,000 Hz (twenty thousand sound waves per second.) Try it out! There are lots of home hearing tests online where you can test your own hearing range.

Every day we hear sounds with a full range of pitch and loudness. Pitch and loudness are two different signals and this mixture of different signals is challenging for your ears to interpret. Usually when we need to adjust a sensory tissue's sensitivity, we simply add more receptors, like in our fingertips and skin where "tuning" the sensitivity is achieved by simply packing in more (or fewer) Piezo receptors. In the ear this won't work. Putting more receptors in the ear might make it more sensitive to quiet sounds, but it still won't tell your brain what pitch your ear was detecting. So how do our hearing receptors detect the difference between sound waves with such a wide range of frequencies? Clearly there is a solution to this problem, otherwise we would only be able to hear a single tone, no matter what frequencies our ears were being exposed to.

One guess might be that there are many slightly different receptors in our ears, each one sensitive to a different frequency.

While this could work, it would also require hundreds or thousands of different receptors to span our hearing range and would greatly complicate the system. Another guess might be that the sound receptors can physically open along a range (from slightly open to fully open) to adjust how many ions are coming through and somehow different ion flows could correspond to pitch. Again, good guess, but the fact is, most (or perhaps all) channel proteins can only exist in either the open or closed state. There is no in-between state where they are just a little open, or a little closed. This "all-or-none" property prohibits an individual channel from being "tuned" simply by increasing or decreasing the size of the pore. So, the question remains, how is an all-or-none receptor channel protein able to respond to and discriminate this very large frequency range? The answer is that the actual channel protein can't. Instead, the inner ear has some spectacular modifications that enable us to detect different pitches with a single type of receptor protein.

Parts of the ear

To understand the answer to the frequency question I just asked, you need to know how the ear works. You likely have heard about the eardrum, which is part of the outer ear. It vibrates when sound waves run into it and causes the bones in the middle ear to move. As shown in figure 1, the working parts of the middle ear include three tiny bones called the *malleus* (the hammer), *incus* (the anvil), and *stapes* (the stirrup). The hammer physically touches the eardrum, so when the eardrum vibrates, the hammer vibrates too. Since the hammer, anvil, and stirrup are all connected by flexible attachments, vibrations of the air are mechanically transmitted from the eardrum all the way to the inner ear. Amazingly, the human eardrum needs to move less than a picometer to make a sound audible! A picometer is one-billionth the size of a millimeter, a distance that is even smaller than the size of a single hydrogen atom! Amazing! The bones of the middle ear are the smallest bones in your body at only a couple of millimeters long. It's important for the eardrum to be relatively large compared to these tiny bones since these bones need to move in response to eardrum vibrations. If these bones were larger, they would move less and transmit less energy (and less information about the ear drum vibrations) to the inner ear.

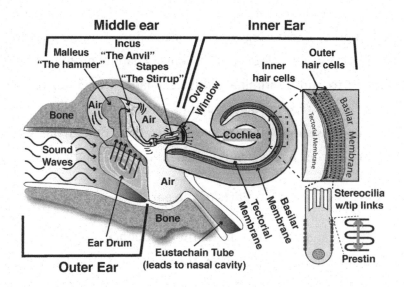

Figure 1. Simplified Anatomy of the human ear.

This arrangement of tiny bones in the middle ear seems rather complex, but the complexity actually serves several really important functions. For instance, the hammer and anvil are shaped in such a way as to give them the mechanical advantage of a lever and can actually amplify the vibrations coming from the eardrum. The eardrum has a larger surface than the end of the stirrup, so the intensity of vibrations ultimately transmitted to the inner ear are amplified about twenty times more than they would be without this arrangement.

The end of the stirrup makes contact with and covers the *oval window*, the first part of the inner ear, which is also the opening into the *cochlea*. As the stirrup bangs on the oval window, which is actually a thin membrane, it sets the fluid inside the cochlea to vibrating. Why is the cochlea filled with fluid? It's probably because vibrations move better through liquid than through air. By pushing vibrations through fluid, more energy can be transmitted to the receptors deep inside, and ultimately the receptors convert the mechanical vibrations into electrical signals.

Despite the benefit of waves moving more easily through liquid, a fluid-filled cochlea also creates a problem; sound waves are very inefficient at passing from air into water. Think about what it sounds like when you are underwater and somebody above water yells at you. You can barely hear them, right? It doesn't work very well because the sound waves bounce off the surface of the water. This problem of air-to-water transmission is solved by the stirrup physically banging on the cochlea to start fluid vibrations in the inner ear. In this way, the sound transmission into the cochlea is more like banging rocks together underwater rather than trying to yell to an underwater swimmer from the edge of the pool. Actually, if the stirrup didn't bang on the oval window and vibrate the fluid inside the cochlea, we wouldn't be able to hear in the air! Amazingly, scientists believe that the eardrum and middle ear structures evolved independently in mammals, reptiles, and birds [44]. Another nice example of convergent evolution. Some evolutionary adaptations are so important that they evolve multiple times; you just can't keep a good adaptation down!

As shown in figure 1, the cochlea of the inner ear looks like a curled-up snail shell. If a human cochlea were laid out flat it would be a little more than an inch long. Inside, a long, wedge-shaped filament called the *basilar membrane* runs down the middle of the cochlea for its entire length. Remember our initial question about how a single receptor can discriminate between different pitches? The basilar membrane is the answer.

The basilar membrane

A convenient and simplified way to imagine how the basilar membrane works is to imagine a harp. The strings of the harp are short and tight on one end, and longer and less tight on the other end. If you pluck the short tight string you get a high pitch and if you pluck the longer and less tight string you get a low pitch. What you might not know is that those same strings can actually be made to vibrate if they are hit with a sound wave that closely matches the frequency the strings usually oscillate at. It's a two-way street; the string vibrations can make sound waves, and sound waves can also make the strings vibrate. Now, for the really important part. Just like the harp, the entire basilar membrane doesn't vibrate at the same time. Only discrete sections vibrate, just like plucking only a few strings at a time. To be perfectly honest, most scientists who study how hearing works would not agree with this analogy, but the harp analogy is simple to grasp and accurate enough for a good understanding of the basilar membrane. Most scientists feel that the cochlea works in more of a standing wave fashion, where the peak of the wave happens in a discrete part of the basilar membrane such that only one section vibrates, just like a single harp string might vibrate. This debate has been going on for over a hundred years, so no matter how it actually works, what's really important is that individual sounds will vibrate isolated parts of the basilar membrane depending on the pitch of the sound.

Imagine taking that harp, laying it on its side and curling it up inside the cochlea. The short tight strings that make the highest notes would be close to the oval window and the long, less tight strings would be located at the far end of the cochlea. This is essentially how the basilar membrane works except that the

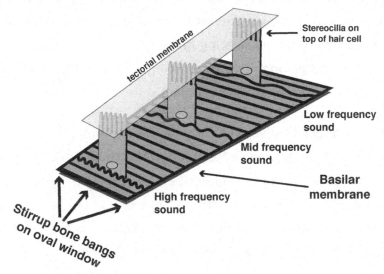

Figure 2: simplified model of the basilar membrane. *The tissues of the wedge-shaped basilar membrane are tight close to the oval window and loose further from the oval window. Individual sound frequencies are separated on the basilar membrane based on which part of the membrane they resonate with. Hair cells are located on top of the basilar membrane and vibrate in response to basilar membrane vibrations. Stereocilia on top of hair cells rub against the tectorial membrane to help activate mechanoreceptors located at the peak of the stereocilia.*

basilar membrane is not made of strings. Rather, it consists of a single sheet of tissue, a membrane, that runs all the way down the cochlea. The membrane starts out very thin and tight near the oval window and gradually gets fatter and less tight towards the end of the cochlea. Just like the harp, the whole basilar membrane does not vibrate at the same time. Instead, individual discrete sections vibrate if they are hit with waves of the correct resonating frequency. Since the very beginning of the basilar membrane is tightly stretched, it begins vibrating when it encounters sound waves with very high frequencies. Conversely, the distant end of the membrane is loosely stretched and only vibrates when hit by low frequency waves. All sounds within our hearing range will resonate somewhere along the basilar membrane according to their frequencies and cause that section to begin vibrating.

Now consider this: very few sounds we encounter in our lives are composed of a single frequency. Most sounds are complex and consist of multiple frequencies mixed together, like a piano chord that is made by pressing multiple keys at the same time. Luckily, the basilar membrane can also deal with these complex sounds by simply allowing each frequency within the complex sound to vibrate its own section within the basilar membrane.

I think that this is simply amazing! This elegant solution of isolating individual frequencies along the basilar membrane is the first part of how we can detect different pitches. And it works quite well. Think about the subtle difference in sound that two adjacent keys on a piano make when struck. Think of that complex sound made when a chord is struck on the piano. Most people can easily detect these subtle differences thanks to their basilar membrane. While this is all very interesting, it is crucial to understand that we do not directly hear the vibrations of the basilar membrane. In order to detect the vibrations of the basilar membrane, we still need a receptor protein. We need a receptor that can translate vibrational energy within the basilar membrane into the electrical energy that the brain can understand.

Hair cells

The receptor is found within the spectacular *hair cells* that grow on top of the basilar membrane. These cells get their names from the hair-like extensions called *stereocilia* that extend from the tops of the cells. Each human cochlea has about sixteen thousand hair cells that come in two varieties, inner and outer hair cells, each with different functions. The inner hair cells are connected to many nerves that travel back to the brain. Because of this *innervation*, inner hair cells are the main cells that send information back to the brain. The outer hair cells have only a few nerves connected to them and are not as important in relaying information to the brain. Instead, these outer hair cells are signal amplifiers. Working together, the inner and outer hair cells are both critical for our sense of hearing and damage to either will cause hearing loss.

Since hair cells grow on top of the basilar membrane, the hair cells themselves vibrate a little when the basilar membrane vibrates. But remember, the entire basilar membrane does not vibrate at the same time. Only isolated regions of the basilar membrane vibrate at a time depending on frequency. Putting this all together, any given frequency will vibrate a small section of the basilar membrane, which in turn vibrates just a few hair cells. Since the sound receptor is located in the hair cells, only those hair cells that are vibrating will send messages to the brain. Since each inner hair cell is individually connected to the brain, the brain is able to decode these vibrations and assign pitch to the sound that we then "hear".

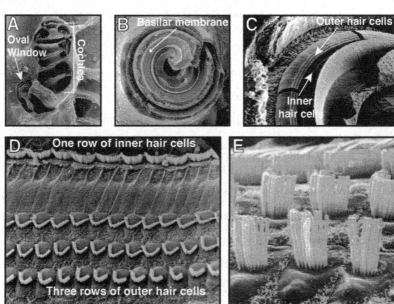

Figure 3: Scanning electron microscopy tour of inner ear structures. A. A side view picture of a cochlea from a chinchilla. The bony outer wall of the cochlea has been removed so we can see inside. The arrow shows the oval window mostly covered by the stirrup, which bangs on the oval window to send vibrations through the perilymph inside. B. A top view of the cochlea. At this magnification, the rows of stereocilia projecting from the surface basilar membrane are just barely visible (arrows). C. Higher magnification of the basilar membrane with stereocilia projecting from the surface. A single row of inner hair cells and a triple row of outer hair cells are indicated with white arrows. D. High magnification view of a row of inner hair cells (top) and three rows of outer hair cells (bottom). E. Very high magnification view of stereocilia. All images used with permission from Robert Harrison (https://lab.research.sickkids.ca/harrison/).

A hairy situation: Converting vibrations to electrical signals

Let's start with the inner hair cells as these are somewhat less complex than the outer hair cells. As I just said, based on the fact that inner hair cells have greater innervation, these cells are the primary cells responsible for relaying information to the brain. Inner hair cells detect vibration through a mechanoreceptor mechanism similar to the Piezo receptor protein. In fact, when Piezo was first discovered, scientists immediately started experiments to determine if Piezo was responsible for hearing. Alas, despite the similarities, the Piezo receptors have not been convicted as being responsible for hearing [45]. In fact, as of the writing of this book, a specific "hearing receptor" has not yet been fully described but we do know a little bit about it.

What we do know is that two proteins called *transmembrane channel-like proteins 1 and 2*, or TMC1 and TMC2, appear to be essential parts of a larger, multi-part mechanoreceptor protein complex in hair cells [46]. TMC1 and TMC2 are located in the extreme tips of stereocilia, are required for hearing, and are thought to form the actual channel of this multi-protein receptor complex [47]. Although we don't understand how this channel works yet, it is interesting that it appears to be composed of **multiple, different proteins**. In comparison, the other receptor channel proteins we have encountered so far like TRPV1, $Na_v1.7$, and Piezo have all been made by large complexes containing 3-4 subunits of the same protein.

The TMC1 and 2 proteins are located very specifically at the very tip of the hair cells, in the stereocilia. This is no accident. The tip is probably the location that wiggles the most when hair cells vibrate. Moreover, the tips of individual stereocilia are connected through *tip links*, which are like ropes stretching between stereocilia that help all the stereocilia on a hair cell to wiggle at the same time. You can see tip links in figure 5 below. It has been speculated that the TMC proteins and the rest of the yet-to-be-discovered mechanoreceptor are directly connected to these tip links such that pulling on the tip links helps to open mechanoreceptors by yanking on them directly.

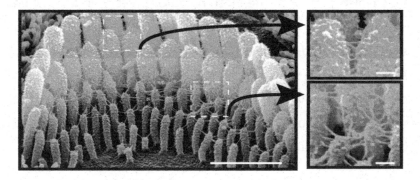

Figure 5: Stereocilia tip links. *Tip links connect individual stereocilia to each other and help to activate mechanoreceptors located on the stereocilia. Boxed insets are further magnified to the right to more clearly show tip links. Images are open access and were taken from [48].*

Finally, localization of mechanoreceptors to the tips of the stereocilia also puts them into contact with the *tectorial membrane*. This membrane covers the hair cells – the very tip of each stereocilia is imbedded into the tectorial membrane. Figure 6 shows a tectorial membrane that has been disconnected and rolled back from the stereocilia. If you look carefully, you can see the holes on the underside of the tectorial membrane where the stereocilia were "stuck" into it. This appears to be the job of the tectorial membrane; to provide a surface for mechanoreceptors to rub against, thus increasing friction against the mechanoreceptor and increasing hair cell sensitivity. Once these mechanoreceptors are activated, they open their channels, the hair cells depolarize, and eventually a nervous impulse travels back to the brain following the same process we saw before for nervous system conduction.

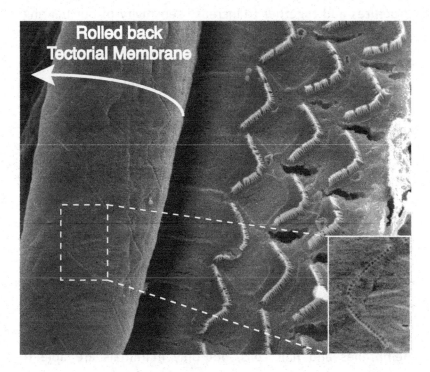

Figure 6. Stereocilia impressions on tectorial membrane. If you look very closely, you can see indentations in the tectorial membrane that has been rolled back from the surface of the hair cells. In the lower right corner, I have enhanced and enlarged a part of the tectorial membrane to better visualize the stereocilia indentations. Image courtesy of Robert Harrison. https://lab.research.sickkids.ca/harrison/

Let's just pause for a moment to appreciate this system. The hair cells are uniquely suited to serve as receptors for vibrational energy. Hair cells vibrate when the basilar membrane below them vibrates, and since the basilar membrane does not all vibrate at the same time, only a few hair cells will vibrate for a given frequency. Other modifications that make this system work are the stereocilia, tip links, receptor localization to the stereocilia tip, and the tectorial membrane. All of these structures add together to increase sensitivity of our hearing. When you take all of these parts and think about them as a whole, I hope you are beginning to appreciate that your sense of hearing is a phenomenal system. That said, I think you will appreciate your hearing even more once you learn how outer hair cells work.

Turn it up to 11: How outer hair cells amplify vibrations

As we just discussed, the job of the inner hair cells is to send information back to the brain. While vibrations of the basilar membrane do vibrate inner hair cells, this is actually not sufficient by itself to cause the inner hair cells to depolarize or to give mammalian ears their sensitivity and ability to detect higher pitches. Indeed, many non-mammals also have basilar membranes and hair cells, but only mammals are able to perceive higher frequency sounds above ~5,000 Hz. Other non-mammals like reptiles, birds, and fish can only hear relatively low frequency sounds less than ~5,000 Hz. One of the major evolutionary advances that mammals have but non-mammals do not have is a second set of hair cells, the outer hair cells. These cells provide an amplifier function that increases hearing sensitivity and gives us the ability to hear high frequency sounds.

Inner hair cells and outer hair cells have many things in common. Just like the inner hair cells, the outer hair cells are also sitting on the basilar membrane. When the correct sound frequency vibrates a section of the basilar membrane, the outer hair cells also wiggle and rub their stereocilia against the tectorial membrane. Outer hair cells also have the TMC1 and TMC2 mechanoreceptors that cause the cells to depolarize in response to basilar membrane vibration, but this is where the similarity ends. Outer hair cells are not as well connected to the nervous system and generally do not send impulses back to the brain. Instead, the outer hair cells use the energy associated with depolarization to amplify the vibrations of the basilar membrane. The amplifier function of outer hair cells is due to a marvelous protein called *prestin*.

A little motor in your ear

Prestin is a *voltage-sensitive motor protein* [49]. That's a mouthful so let's break it down. You already know about voltage-sensitive proteins like $Na_v1.7$; they change conformation in response to changing transmembrane voltages. *Motor proteins* do mechanical work, or exert forces, within a cell. So, a voltage-sensitive motor protein like prestin, does mechanical work in response to changes in voltage.

The prestin protein is found nowhere else in your body besides the edges of outer hair cells. The protein weaves in and out of the cell membrane multiple times and it also binds to negatively charged chloride ions (figure 7). Based on this information, a model has been proposed to explain how prestin might work. It goes something like this: recall that resting cells are negatively charged on the inside. Since like-charges repel each other, negatively charged ions like chloride (Cl^-) try to escape from the negative environment of the cell. The current hypothesis is that these chloride ions take refuge **inside of the prestin protein**. When they do, all those chloride ions cramming into prestin forces the protein to spread out and elongate within the membrane.

Think of prestin sort of like an accordion imbedded into the cell membrane. When the chloride ions enter into prestin they force the accordion into the stretched-out shape. Working together, millions of prestin molecules cause the cell to stand up a little taller. Everything changes when the cell depolarizes. The cell interior becomes positively charged for a few milliseconds and, during this crucial period, the negatively charged chloride ions within prestin are attracted to the more positive charge inside the cell. The chloride ions are pulled out of prestin and the protein is allowed to briefly collapse into a smaller structure, sort of like how the accordion would rebound to its normal shape if allowed to relax. When this happens, the segments of prestin that weave through the cell membrane collapse towards each other and pull their attached segments of cell membrane closer together. This makes the entire cell sort of scrunch up a little. When the cell repolarizes and the intracellular negative charge is restored, the chloride ions are repelled out of the cell and again retreat back into prestin. This causes prestin to expand and spread the cell membrane out again, and the cell returns to its full-upright "unscrunched" position. Keep in mind that nobody has actually seen this yet, it is only our best guess, or model, for how Prestin works.

Even though an individual prestin molecule is only a few nanometers long, there are estimated to be ten million prestin molecules in each outer hair cell. Every time hair cells depolarize, millions of prestin molecules in the sides of the cell scrunch up and pull the membrane in, causing the cell to shrink a little. Then, as prestin becomes stretched out again, the cell returns to its original size.

Outer hair cells really do this! Grab your device and do a little Google search for "dancing hair cell". You should find a video of an isolated outer hair cell under a microscope. In this

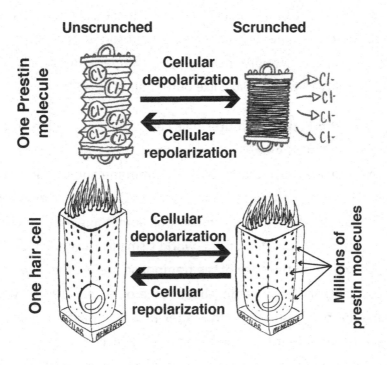

Figure 7: Model for prestin function in outer hair cells. In resting (polarized) cells, negatively charged chloride ions situate themselves inside prestin. During depolarization, the cell interior becomes more positive and pulls the negatively charged chloride ions out of prestin. Without chloride ions inside, prestin collapses to a more compact scrunched structure. The process is reversed after cellular repolarization and reestablishment of the resting transmembrane potential. Millions of prestin molecules in the hair cells cooperate to make hair cells lengthen and shorten in response to each cellular depolarization.

video, instead of being stimulated by vibration, the cell has been impaled with a fine glass needle that can rapidly change voltage and cause the cell to artificially depolarize. The cell is contracting, or "dancing," to the tune of *Rock Around the Clock* by Bill Haley & His Comets. We know that prestin is responsible for this because if it is deleted from these cells they cannot dance to the tune [50]. Conversely, if prestin is introduced into cells that don't usually make it, prestin imparts these cells with the ability to contract just like outer hair cells [49]!

So prestin makes the outer hair cells dance in rhythm to the vibrations of the basilar membrane. As they do this, the hair cells pull and push on the basilar membrane and this amplifies the vibrations within that little region of the basilar membrane. I always imagine that this must look something like the outer hair cells are jumping on the world's smallest trampoline! The bystanders on the edge of the trampoline are like the inner hair cells and they are getting bounced around because of what the jumpers are doing. No matter how you might imagine it, the result is that the innervated inner hair cells receive about a hundred times more vibrational energy from the basilar membrane than they would without the outer hair cells. This additional vibrational energy is required for the inner hair cells to fully depolarize. Without the amplifier function of outer hair cells, humans cannot hear.

Before we leave prestin behind, let's consider what this unique protein is really doing. We have looked at heat-sensitive and ligand-gated channels like TRPV1, voltage-gated channels like $Na_v1.7$ and $Ca_v1.3$, and even the pressure-sensitive Piezo channels. These channels are all interesting in that they can convert a signal of some sort into a cellular depolarization.

Prestin is quite different in that it **converts outer hair cell depolarization into mechanical action**. It's like the reverse of a Piezo channel and the only voltage-gated motor protein that we know of. A truly amazing bit of molecular biology that we are only just now starting to appreciate.

When good hair cells go bad

Unfortunately, the wear-and-tear of aging, loud noises, and even some *ototoxic drugs* that specifically damage hair cells, all contribute to our hearing "wearing out". This is usually accompanied by loss of the stereocilia on hair cells. Figure 8 shows damage to hair cells. Panel A shows what happens to hair cells that have been exposed to loud, sustained noise. Notice that several clusters of stereocilia are missing from the outer hair cells, while the inner hair cells look normal. For unknown reasons, outer hair cells are more susceptible to loud sounds, while the inner hair cells are more resistant. This loss of outer hair cells is a testament to the importance of the outer hair cells amplifier function. Once these cells are damaged, hearing

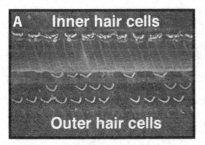
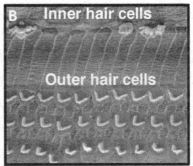

Figure 8. Damage to stereocilia causing hearing loss. In panel A, loud music has damaged the outer hair cells. In panel B, the ototoxic drug carboplatin has damaged the inner hair cells. Image courtesy of Robert Harrison. https://lab.research.sickkids.ca/harrison/.

is diminished or lost. Understanding how loud noises cause hearing loss is a little easier to understand once you understand how hair cells work. Just imagine what listening to heavy metal music on full blast does to those outer hair cells. It must look like a mini mosh-pit in your ear!

Panel B shows a different sort of damage; the effect that the anti-cancer drug *carboplatin* can have on inner hair cells. About 20% of people treated with carboplatin will experience some hearing loss, but we don't yet know how or why carboplatin specifically targets inner hair cells. Other ototoxic drugs only affect the outer hair cells, but again we don't yet know why. In either case, once hair cells are lost, they cannot be regrown and hearing damage is permanent.

An interesting fact of hearing loss is that most people with hearing damage caused by loud sounds first lose the ability to hear high pitches. This is because the part of the basilar membrane that detects high pitches is closest to the oval window where vibrations first enter the cochlea. Hair cells on this part of the basilar membrane experience the most intense vibrations and are damaged first.

Once damaged, hair cells cannot be regrown (although people are currently working on doing just that [51]). There are some options for individuals with hearing loss, though. Traditional hearing aids work as a simple amplifier to increase volume, and therefore vibrational energy, in the cochlea. But this is only useful if a person still has some hair cells remaining.

In comparison, some people are born deaf. The most common cause of genetic deafness is a mutation in a gene called *GJB2* that makes a protein called a *connexin*. Individuals with this

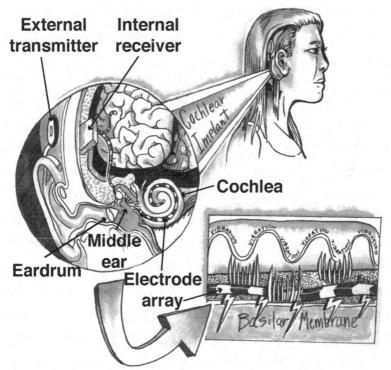

Figure 9: Cochlear implant. *A cochlear implant is a bundle of thin wires (electrode array) that are threaded into the cochlea and make contact with the basilar membrane. An external microphone captures sounds, sends the informtaion to the external transmitter, which wirelessly transmits to the internal receiver which is implanted under the skin. The internal receiver sends electrical impulses down the electrode array which stimulates specific regions of the auditory nerves thus bypassing damaged or non-functional hair cells.*

mutation are "profoundly deaf" and cannot hear anything at all. Connexin proteins are involved in making *gap junctions*, which are essentially "tunnels" that connect adjacent cells and allow ions to pass back and forth between them. Mutation of the GJB2 gene seems to interfere with the normal potassium distributions in the inner hair cells. The cells lose the ability to depolarize so they cannot send impulses through the auditory nerves. The auditory nerves are still present though and if they are artificially stimulated, they still send information back to the brain.

This is what a cochlear implant does. It bypasses the hair cells and directly stimulates the auditory nerves. The implant has a long thin wire that is threaded into the cochlea where it makes contact with hair cells and the auditory nerves. An external transmitter detects sounds and broadcasts this information to an receiver that is implanted into the skull and attached to the wire. The receiver detects sounds, then activates the wire by delivering a small voltage to individual regions of the basilar membrane, artificially activating the auditory nerves in that region. Different segments of the wire are activated to stimulate the auditory nerves in the regions of high or low pitch so people with the implant can discriminate between different pitches. This relatively new technology has deeply changed the lives of the profoundly deaf people who chose to use it. Imagine how wonderful it would be for a deaf adult to hear the voice of their child for the first time!

Ultrasonic hearing

As I mentioned before, humans can detect a range of frequencies from about 20 Hz to 20,000 Hz. Some animals can detect *infrasonic sounds*, or frequencies below about 20 Hz. Other animals can detect *ultrasonic sound*, or frequencies above 20,000 Hz. For instance, dogs can detect frequencies up to about 50,000 Hz. Dog whistles usually send out sound waves at frequencies about 25,000 to 50,000 Hz, which is why dogs can hear those whistles but we can't.

Some animals have an even greater range of hearing, and

some can actually manipulate and put these ultrasonic sounds to work. For instance, bats not only sense ultrasonic frequencies, they also commonly produce ultrasonic sounds in the range of 20,000 to 100,000 Hz for hunting and detecting their environments. *Echolocation* is when an animal emits ultrasonic sound waves, then detects the waves that bounce back at them. Bats, dolphins, and whales can all do this. It's a real "superpower" that allows these animals to work in the complete dark.

Some work towards understanding how echolocation works has been done. One interesting finding is that the cochlea from echolocating bats is longer and has more turns compared to the cochlea from non-echolocating bats [52]. This suggests that the basilar membrane may be tuned to a higher pitch in echolocating bats. In addition, several years ago, two independent groups wanted to understand the cell and molecular biology of ultrasonic hearing. They found that a newer form of prestin, which likely evolved from a genetic mutation in the prestin gene at some point, seems to be critical for dolphins and bats to be able to hear ultrasonic frequencies [53, 54]. The current belief is that the mutant form of prestin simply causes the hair cells to have a greater baseline tension which makes them more sensitive to higher frequencies, much like the tightest guitar string produces the highest pitches and the tightest end of the basilar membrane is sensitive to the highest frequencies [55].

While it is amazing that this "superpower" might be due to a single protein, it's also amazing that echolocation seems to be yet another example of convergent evolution. The most interesting thing here is that, not only did bats and dolphins independently develop echolocation, but they did it in precisely the same way; mutations in the same gene affected the **exact same amino acids of the prestin protein**, which caused the **forms and functions of the protein in each individual species to change in the exact same ways.** The chance of both bats and dolphins evolving in this way is very low, and the fact that it happened like this is truly remarkable. It's an excellent example of what random mutation and evolution can achieve if given enough time.

Do you hear what I see?

I cannot move on from this topic without drawing your attention to a few more fascinating things about hearing. These are not necessarily related to the molecular biology of hearing, nonetheless they are immensely interesting. First, have you ever wondered what it would feel like to use echolocation? What would the world "look" like as bouncing sound waves? Bats and dolphins can detect moths or fish in the complete darkness and catch them, even as the moths and fish are trying to evade them. So clearly echolocation gives information about direction and speed. But do bats or dolphins physically "see" the moth or fish the same way we do? Does the echolocation system give these animals a mental picture of their world that is similar to what we perceive with our eyes?

We don't know, but there are some blind humans who can use echolocation to navigate their world. They click their tongues and listen for the sounds that return after bouncing off surrounding objects. People with this ability describe the world they perceive as being three dimensional and rich, and they are able to do things that require navigating through space, like riding bikes, taking walks, and cooking. Daniel Kish gives a nice

TED Talk called *How I Use Sonar to Navigate the World.* Check it out if you're interested.

Bats use echolocation to catch bugs in the pitch dark. In some cases, however, the bugs aren't taking this abuse anymore and have evolved defense strategies. A full-blown evolutionary contest is being played out between bats and moths. Some moths can actually detect the ultrasonic emissions of bats and will evade the bat by simply folding their wings up and falling from the sky. The tiger moth takes a more aggressive tactic. When it hears a bat's ultrasonic frequencies it emits an ultrasonic burst of its own. The tiger moth's ultrasonics seem to effectively jam the bat's radar, confusing the bat and making it miss its target [56]. When scientists silenced tiger moths by removing their ultrasonic generator organ, the bats caught more silenced moths than unaltered moths. There are some really captivating videos online that show the bats catching or missing moths. Do a search for "moth jamming bats on National Geographic" and you should find the videos. Some moths also seem to have evolved sound-absorbing scales on their wings that absorb the bat's ultrasonic bursts, instead of rebounding them back to the bat [57]. These scales appear to provide the moth with a "stealth mode" to avoid detection by bats.

Chapter summary

Before we move on to the next chapter, lets summarize what we learned about the cell and molecular biology of hearing. We learned that sound waves are efficiently transmitted into the cochlea by the stirrup bone thumping against the oval window. These thumps set the fluid in the cochlea to vibrating, and different vibrational frequencies resonate with a corresponding section of the basilar membrane depending on the membrane's tightness in that section. As the basilar membrane begins to vibrate, outer hair cells, with the help of the voltage-regulated prestin protein, amplify the vibrations. Inner hair cells detect the vibration with the mechanoreceptors at the tips of their stereocilia, undergo cellular depolarization, and send a nerve impulse back to the brain. Our sense of hearing is certainly one of our most complicated senses since it involves so many conversions of energy. It starts with the conversion of sound waves in the air into mechanical energy in the middle ear, into fluid waves in the cochlea, into vibrational energy on the basilar membrane, then back into mechanical energy as the tectorial membrane "rubs" on the tips of the hair cells, then, finally, into electrical signals. Our sense of vision which we will study next, is also complex, but it doesn't have as many energy conversions.

Chapter 7: Vision

Claude Monet is one of the most well-known impressionistic painters of the 19th century. It is believed that later in life, after eye surgery, Monet was able to see into the ultraviolet spectrum and this might help explain his choice of blue-tinted colors when painting his famous water lilly paintings.

As humans, most of us value our sense of vision more than our other senses. This is understandable given that we receive more information through our eyes than through our other senses. And yet, do you know how your eyes work? Most of us remember that light is focused through a lens in our eye onto the retina. Most of us also know that the retina has rods and cones that detect the light and send information back to the brain. For most people, that is where the knowledge ends. Did you ever stop to wonder what exactly is happening inside those rod and cone cells when light shines on them? For that matter, have you ever been curious how you can detect different colors of light? Similar to our ears, our eyes need to convert one kind of energy (light) into another kind of energy (electrical) in order for our brains to understand the messages it is receiving from our eyes. As in our discussion of touch and hearing, vision also depends on receptor proteins (to detect light) and nerve fibers (to send the electrical impulse to the brain). Since light is very different than either pressure or sound vibrations, the receptors are quite different. They are unlike anything we have talked about so far.

Let's talk about light

Any discussion about vision should start with a discussion about the nature of light. A physicist would start by explaining the idea of *wave-particle duality,* or the idea that light acts like both a wave and a particle. Physicists have argued for decades about whether light is a particle or a wave and, at this point, we have pretty much settled on describing it as both, depending on the situation. We are not going to settle this debate here, but let's agree that light is a form of *electromagnetic radiation.*

For our purposes we are going to boil it down to this: all electromagnetic radiation is made of *photons.* Photons are single "units" of electromagnetic radiation that have slightly different amounts of energy, slightly different wavelengths, and no mass. This means that there is a full spectrum of photons flying around everywhere, all the time, each with a different energy and wavelength. We cannot sense most of the electromagnetic spectrum, but it is always there, just beyond our perception. We call the little bit that we can detect *light.* The light we see is composed of different colors. "ROY G. BIV" is an acronym often taught in school to help people remember the colors of the visible spectrum: Red, Orange, Yellow, Green, Blue, Indigo, and Violet. The red color at one end of the rainbow is composed of photons with longer wavelengths and lower energies, while the violet end of the rainbow is composed of photons with shorter wavelengths and higher energies. Light waves, or photons, bounce around and get reflected off things, so if we can detect these photons, we can form an image of the world around us.

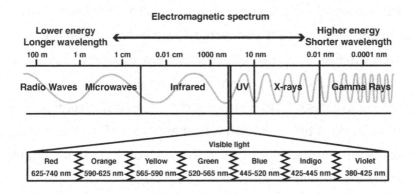

Figure 1: The electromagnetic spectrum. Photons of energy in the electromagnetic spectrum are differentiated by wavelength and how much energy they contain. Photons with wavelengths between 625 and 425 nm can be detected by our eyes and appear as visible light.

Evolution of the eye

The story of the eye is one of evolution's greatest hits. Complex eyes didn't just appear on Earth one day, instead they started as simple eyes and had many intermediate steps that got progressively more and more complex. We know this because examples of even the most rudimentary eyes are found in animals on Earth today, and we can study the evolutionary relationships between these organisms by comparing their DNA sequences. Many of these intermediate eyes still exist because they are evolutionary "milestones" that were particularly successful for an organism and continue to be useful today. Interestingly, as we saw with bioelectricity and hearing, the evolution of eyes is another example of convergent evolution. A most striking example is the similarity between the eyes of an octopus and our eyes; they evolved completely independently yet have very similar final structures.

The simplest eyes we know about are little more than clusters of light-sensitive cells connected to a simple nervous system. *Eyespots* like these can tell an organism if its environment is light or dark, but they cannot produce any kind of image or tell the organism from which direction the light is coming. The experience of this sensation is probably something like looking at the sun through your closed eyelids; you can tell it's bright, but really nothing else.

The next step in the evolution of the eye is thought to be the *eyecup* in which the light-sensitive cells are pulled into a recess within the organism's body. The evolutionary advantage of the eyecup is that, depending on where the light is coming from, one side of the cup will be more shaded and the other side will be illuminated, so the animal's brain can interpret from which direction the light is coming.

Moving forward, the edges of the eyecup draw closer together, resulting in an eye that works like a *pinhole camera*, only allowing a very small amount of light to pass through it. The pinhole is a major evolutionary step forward since this type of eye can produce a rough image of what the animal is looking at. You may have made a pinhole camera to observe a solar eclipse, or maybe you make a pinhole with your fingertips to improve your vision when you forget your glasses? The image isn't exactly perfect, but at least it's better than having no glasses at all.

Eventually, the pinhole gets covered with a layer of clear cells and a *lens* develops, which helps to focus the light waves entering the eye.

This is a neat and tidy evolutionary path, but evolution is not

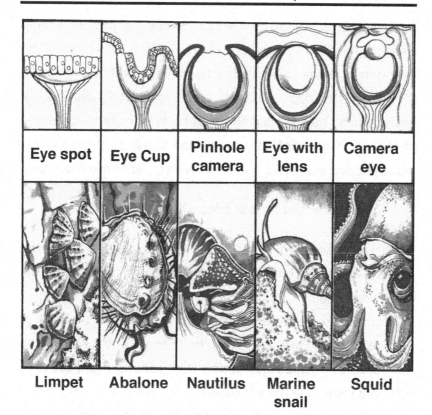

Eye spot	Eye Cup	Pinhole camera	Eye with lens	Camera eye
Limpet	Abalone	Nautilus	Marine snail	Squid

Figure 2. Different types of eyes found in nature. The first eyes on Earth were simple eye spots that evolved over time into more complex eyes.

always so linear. An impressive example of this is the scallop, which sports a marvelous set of up to two hundred eyes. If you look up a picture you can see dozens of them peeking out from under the edge of the scallop's shell. Rather than using a lens to bend light, like our eyes do, scallop eyes use tiny mirrors to reflect light onto their retina. This is possible because molecules of guanine, in addition to building DNA, can also build shiny crystals, and these crystals are really good at reflecting light. Some animals, like carp, neon tetras, zebrafish, chameleons, and some plankton, use guanine crystals to produce iridescence or to change color [58]. Scallops use them to reflect light in their eyes. Each scallop eye contains millions of tiny and almost perfectly flat guanine crystals arranged in a bowl shape to reflect light onto a retina [59, 60]. Exactly how the scallops can make these intricate arrays of crystals is not completely understood, but scientists believe that the molecules of guanine crystalize inside individual cells that crawl around and arrange themselves into a checkerboard-like pattern to reflect light onto the retina.

The retina: The land of happy little rods and cones

Most people seem to have a pretty good understanding of the basic way that light enters an eye and ends up as an image in the brain, so I'm going to go through this pretty fast. Eyes like yours and mine work by focusing light onto the retina in the back of the eye, basically like a camera. The retina is a thin membrane composed of cells, both light-detecting cells and support cells, that hold the whole thing together. Light that reaches the retina is detected by two kinds of cells: *rods*, which give us black and white vision, and *cones*, which give us color vision. Each individual cone is connected to a single nerve fiber that travels back to the brain. This "one-cone-one-nerve" arrangement means that the brain knows exactly what information is coming from which cone. As a result, "cone vision" is very sharp and precise. On the other hand, multiple rods often connect to the same nerve fiber. This "many-rods-one-nerve" arrangement means that the brain is never quite sure from which specific rod a bit of visual information came, and as a result, "rod vision" is less sharp than "cone vision". This is basically what most people already know about how our eyes work. Compared to the basic function of the ear, the basic function of the eye is relatively simple. That is, until you get into its cell and molecular biology. Then...I'll let you be the judge.

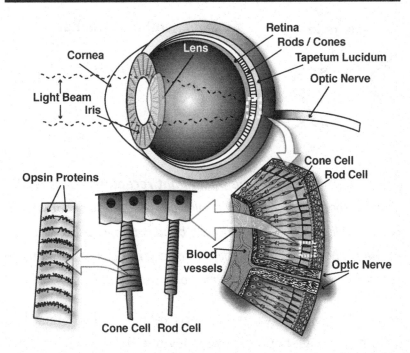

Figure 3. Basic anatomy of mammalian eyes. *Light is focused by the lens onto the retina where the rod and cone cells are located. Rod and cones have many disks in them where the light absorbing opsin proteins are located. The tapetum lucidum is not present in all animals.*

Retinal: The heart of vision

The real heart of our light receptors is a molecule called *retinal*, with the "AL" at the end of its name. Retinal is a slightly modified version of *retinol*, with the "OL" at the end of its name. You probably know retinol better by its more common name, vitamin A. Yes, just as the old carrot myth suggests, our vision is in fact dependent on vitamin A, or a derivative of it anyway. Retinal exists in our rods and cones in several slightly different forms. The two that are the most important are called *11-cis retinal* and *all-trans retinal*, shown in figure 4 on the next page.

11-cis retinal   **All-trans retinal**

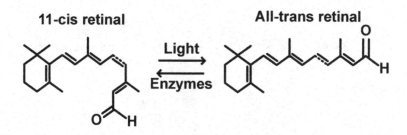

Figure 4: The heart of the opsin light receptors. 11-cis retinal changes shape (is isomerized) to all-trans retinal when it absorbs a photon of light. The dashed double bond is temporarily converted to a single bond that allows the molecule to rotate into the all-trans retinal structure before the double bond is reformed.

In the dark, retinal exists as 11-cis retinal. 11-cis retinal has a ninety-degree bend halfway down the backbone of the molecule. In figure 4, the critical chemical bond is indicated by a dashed line. Notice that the dashed line makes one half of a double bond. Your high school chemistry teacher probably once told you that, unlike single bonds that allow rotation, double bonds are static and the connected atoms cannot rotate. Because of its many double bonds, 11-cis retinal is pretty much stuck in this ninety-degree-kinked position. The "magic" of this molecular is that the critical double bond in 11-cis retinal is able to absorb certain wavelengths of light, and when it does, the double bond is temporarily converted into a single bond. The temporary single bond does allow rotation, and 11-cis retinal converts to the straight (and un-kinked) all-trans retinal form. When a chemical reaction involves no change in chemical formula, just a change in shape, we call this an *isomerization reaction*. The modest isomerization that occurs between 11-cis retinal and all-trans retinal when it absorbs light may not look like much, but it is the heart of vision in all animals that we know of, including you and I. This chemical reaction kicks off the cascade of events that ultimately results in your ability to read this text. It's happening right now in your eyes.

Retinal doesn't work alone, it's just a trigger that initiates the next event. The retinal molecule is "caged" inside of a larger protein called an *opsin*. The opsin protein is a receptor and, like all the other receptors we have talked about so far, it is activated by changing its molecular shape. In figure 5 you can see the actual structure of a cow opsin protein. Right in the center is the "caged" retinal molecule. The "bars" of the cage are seven spring-like protein structures called *alpha-helices*.

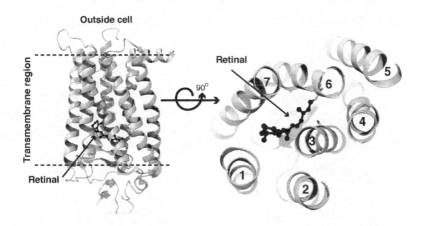

Figure 5. Cow opsin. *The image on the left shows the opsin protein with its captive retinal from the side. The image on the right shows a slice right through the middle of the opsin while looking at it from the top down. From this perspective, you can really see how retinal is caged by the alpha-helices, which are numbered one through seven. Looking at the structure from this perspective it's easy to imagine how the opsin receptor could be distorted by the change in shape that occurs when retinal converts from the 11-cis to the all-trans form; the un-kinked form of retinal simply puts pressure on the bars of the cage and forces the opsin to change its own shape.*

An alpha-helix is just a section of the protein's chain of amino acids that has been twisted into a spiral. In the cow opsin these alpha helices are stuck within the cell membrane (they are transmembrane helices) so that only the top and bottom project from the membrane.

To see this in another more fun way, check out figure 6 below. This is called a "stereoview" of the same opsin protein. Although it just looks like two pictures of an opsin side by side, with a little practice this stereoview will allow you to see the opsin in a new way. It's an old molecular biologists "mind-trick" to see things in three dimensions. To view the stereoview, hold your face about 18 inches from the image and begin to cross your eyes. If you

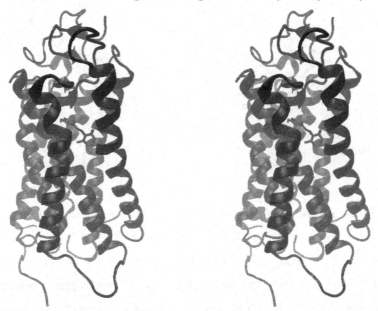

Figure 6: Opsin stereoview. *To see the opsin protein and its caged retinal, hold the image about 18 inches from your eyes and cross them just until you see three images then focus on the middle image. In a few seconds, you should be able to see the 3-dimensional nature of how retinal is caged inside the opsin protein.*

cross your eyes too much, you will see four images because your left eye is looking at the two images independently from your right eye that is also independently looking at the two images. The trick is to cross your eyes only a little bit so that you see three images then really focus on the middle image. In a few seconds, you should be able to get the middle image in focus with no eye strain and you will see the three-dimensionality of the protein including the retinal in its cage. I hope it worked for you. I bet you didn't know your eyes could do that.

Cell signaling in vision: "G" that's really interesting

So far, we know that retinal is caged by opsins and that when retinal absorbs light it changes its shape. Next, I'll describe how this change in shape eventually results in cellular depolarization. This next step is going to really get us into the "molecular weeds". We have discussed many different types of receptors so far and most of these have been channel proteins. Most opsins are not channel proteins and they cannot directly cause cellular depolarization, even with the help of their captive retinal molecules. Instead of directly opening a channel to cause cellular depolarization, opsins convert light into a kind of "cellular message" where intermediate molecules transfer the information about the light to a completely separate channel protein located nearby in the cell membrane. Cellular depolarization and message transmission to the brain is technically caused by molecules that are several steps removed from the opsin protein.

I debated with myself for a long time about whether or not the intermediate steps of message transmission would be an information overload, and whether or not understanding them is even required to understand how vision works. In the end I decided that if you've made it this far you are really interested in knowing how things work in the molecular world. So, here it is, a little bit of the molecular nitty-gritty.

The cellular messaging system involved in vision is called a *G-protein-coupled receptor system*, or GPCR for short. The GPCR system is basically a way for chemical messages to be sent around inside a cell, just like we send emails to friends around the globe. I like to think of these chemical communication pathways as runners doing a relay race where they pass a baton from one runner to the next until the last runner crosses the finish line. This sort of chemical communication goes on all the time inside cells, and the GPCR system is just one of many different communication, or signaling, systems.

Right now, in research labs all over the world, there is a tremendous amount of research going into understanding cell signaling mechanisms since malfunction of these communication pathways often leads to human disease. Think about how difficult it is to communicate with somebody when you don't speak the same language. Or imagine trying to communicate with somebody who cannot speak at all but only yells at the top of their lungs! This is basically what happens when cell signaling mechanisms go awry. The inactivation or hyperactivation of cell signaling mechanisms, like GPCRs, is responsible for many human diseases, including many types of cancer. If you can get a grasp on how GPCRs work, you will have taken a very large step towards understanding some of the most complex inner workings of cells.

As I already mentioned, the opsin proteins are a special type of protein called G-protein-coupled receptors. This is a scientific mouthful, but what it describes is a receptor, such as an opsin, that is attached, or coupled, to **another** protein, called a G-protein. This arrangement is illustrated in figure 7 on the next page. When an opsin is inactive and retinal is in the 11-cis form, a G-protein can bind to the opsin on the piece of protein that is sticking out of the membrane. When 11-cis retinal absorbs a photon, changes shape, and in turn changes the shape of its opsin cage, the opsin's shape, no longer fits the G-protein's shape. The G-protein becomes dislodged from the opsin, and it floats away into the cell.

G-proteins are made up of three separate pieces of protein, or *subunits,* that normally all stick together. As the freed G-protein floats away, one of its subunits is released (again, separation is caused by a change in shape). The subunit floats away and interacts with another protein called *guanylyl cyclase.* Guanylyl cyclase is an enzyme. Its job is to grab molecules of *guanosine triphosphate* (GTP) and modify them into *cyclic guanosine monophosphate* (cGMP). It's not important that you know what GTP and cGMP look like, or even what they do; just understand that the signal (the runner's baton) has been passed from the light to the retinal to the opsin to the G-protein to the guanylyl cyclase, and finally to the cGMP. cGMP is a ligand that **binds to and opens channel proteins in the cell's membrane**, allowing sodium to enter the cell. As we have seen so many times before, as sodium enters the cell, it causes the cell to depolarize and eventually sends a nerve impulse to the brain. Figure 7 shows this process from start to finish and breaks it down into eight steps. Do you see how one step causes the next step to happen? Sort of like those relay runners?

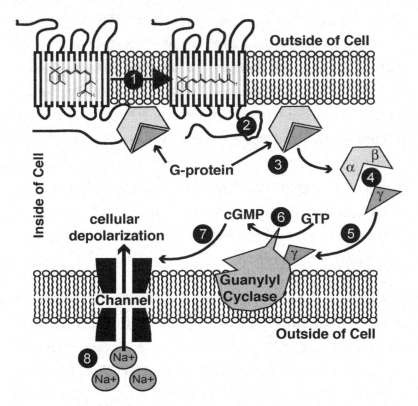

Figure 7. Model of GPCR signaling. 1.) 11-cis retinal molecule in opsin G-protein-coupled receptor (GPCR) absorbs photon and changes to all-trans retinal. 2.) Opsin GPCR changes shape. 3.) Altered GPCR shape causes bound G-protein to fall off receptor. 4.) Free G-protein breaks into two pieces. 5.) Subunit of G-protein binds to guanylyl cyclase enzyme and activates it. 6.) Activated guanylyl cyclase enzyme converts GTP to cGMP. 7.) cGMP binds to and opens channel protein. 8.) Opened channel protein allows sodium ions to enter cell and causes cellular depolarization.

You might be asking yourself right now, "why does this whole process need to be so complex? Why on Earth do we need this whole series of events just to cause cellular depolarization? Wouldn't it be easier to have an opsin protein that is also a channel protein such that it could directly cause cellular depolarization after absorbing some light?"

The simple answer to all this complexity is *signal amplification*. Think of it this way: a little bit of light can activate a single GPCR that eventually activates a single guanylyl cyclase, but, since guanylyl cyclase is an enzyme (a molecular machine), it will continue to change GTP to cGMP **until it is turned off**. During this time, it will churn out as many molecules of cGMP as it possibly can. In this way, a little bit of light can activate a single opsin protein, resulting in the creation of many cGMP molecules, which in turn activate many sodium channels. It's just like the Rube Goldberg/domino example I introduced when we talked about the blood clotting cascade. With this amplification system our eyes are much more sensitive to low levels of light than they would be if our opsin proteins directly opened as a channel. If each opsin acted as its own single channel, only a few sodium ions could enter the cell. The cell might not depolarize enough to send a signal to the brain. The signal amplification allows a very small amount of light to activate many channels and trigger a message to the brain.

This is not the first time we have encountered signal amplification. We also talked about signal amplification when we talked about nerve depolarization and when we talked about prestin function in outer hair cells. It's a common theme in biology that allows us to take small inputs from our environments and make big signals that our brains can respond to.

How to see in Technicolor

I have just described how light energy is converted into electrical energy by rod and cone cells, but this does not fully explain how we can discriminate between different colors. To understand this, you need to remember our discussion about photons. Remember that red light is composed of photons with longer wavelengths and lower energies, and blue light is composed of photons with shorter wavelengths and higher energies. In addition, some chemical bonds like the one in 11-cis retinal can absorb photons, but only if the photon has the correct wavelength. Photons of the wrong wavelength just pass by without being absorbed.

Most animals make more than one kind of opsin protein. Remember that opsins are the "cage" proteins that hold 11-cis retinal captive and change their shape when retinal absorbs a photon. Humans classically make four different types of opsin proteins: one that absorbs green light, one that absorbs red light, one that absorbs blue light, and one for black and white (grayscale) vision. Each different opsin protein holds an identical retinal molecule, but each opsin has a slightly different protein structure, so each different opsin holds its retinal in a slightly different chemical environment. These different environments slightly alter the chemistry of that critical ninety-degree bond in 11-cis retinal and cause the retinal to absorb photons of slightly different wavelengths. Each of your six-million cone cells only makes one variety of the three different opsins. This means that we have "green" cones, "red" cones, and "blue" cones.

The cones are not actually those colors, we just use the color names to describe the range of wavelengths that can be absorbed by the retinal held in the opsins produced by that cone. Red cones absorb light with longer wavelengths from the red end of the rainbow. Blue cones absorb light with shorter wavelengths more towards the blue end of the rainbow. And you guessed it; green cones absorb light best from the middle of the spectrum, between blue and red.

We don't have cone cells for yellow or orange light because we don't need them. We can see yellow because the red and green opsins do not only absorb a single wavelength of light; they have a range of light they can absorb, and these ranges overlap a little. Due to this overlap in wavelength absorption, we can cover the whole spectrum of visible light. If yellow light shines into your eye, both your red and green cones will detect that light, and your brain mixes these signals together to report yellow. It's the same for all the colors you can imagine. Overlapping absorption of different cones adds together in your brain and you interpret a composite color.

How many shades of gray?

But what about grayscale vision? Why do rod cells only detect black and white? Actually, there is no such thing as black and white light. The opsin in rod cells is called *rhodopsin* and it best absorbs light in the green range. We don't see green with our rod cells though since it's our brains that ultimately interpret the information it gets from rods as white.

You might be asking yourself "why do we have rods anyway? Surely color vision is better than grayscale vision, right?" Well, human eyes have about 100 million rods and only about 6 million cones so rods must do something important for us that cones cannot. The advantage of rods is that they are far more sensitive than cones. As a matter of fact, one of the defining characteristics of rods is that they can send a neural signal after a single photon of light activates a single rhodopsin protein. Just one photon! In comparison, a cone cell requires an estimated one hundred to one thousand photon-activated opsins to generate a depolarization. To get a general feel for how many photons this actually is, it is estimated that when you see stars in the night sky, you are detecting roughly between ten and one thousand photons for an individual star. The extreme sensitivity of rods is what gives us the ability to see in low light conditions. Moreover, since rods but not cones work in the dark, things look mostly grayscale in the dark.

How is the amazing sensitivity of rods achieved if rods work using the same GPCR signaling mechanism as cones? A single photon still activates a single retinal molecule, setting guanylyl cyclase into action. How does the rod build up enough cGMP to open enough channels to initiate a nervous impulse? The answer to this question is still being debated, however, as I mentioned before, cones are all connected to a single nerve fiber, and this is not the case for rods. Instead, about three rods are connected to a single nerve fiber and the individual rods work together to make the nerve depolarize.

Another reason that rods are more sensitive than cones is that the guanylyl cyclase enzyme not only needs to be turned on, but it also needs to be turned off. It appears that rod cells turn off their GPCR signaling systems more slowly than cone cells do by keeping their guanylyl cyclase enzymes active for longer periods [61]. This means that rods are able to make more cGMP from a single photon thus amplifying their signals and keeping their sodium channels open longer than cones can. In the end, a smaller stimulus (one photon) is all it takes to depolarize a rod.

One might think that cones are more advanced than rods since they can detect color, but this does not appear to be the case. Evidence suggests that rods may have actually evolved from cones and that this "one-photon activation" was the critical push for the evolution of our rods.

So, we have learned the cell and molecular biology behind sight. We have seen that photon absorption changes the shape of 11-cis retinal and this causes shape changes in opsin proteins and the eventual production of cGMP, which opens channel proteins. We have also learned that color vision is possible because red, green, and blue opsin proteins have slightly different chemistries that cause their caged retinol molecules to absorb different wavelengths of light. Finally, we have learned that cell signaling events are responsible for the signal amplification that increases sensitivity in our eyes and gives our rods more sensitivity than cones. Now that you have an understanding of how your eyes see at the molecular level, let's compare our vision to that of some other creatures.

Vision in other animals

We can see both color and grayscale to suit either high or low light conditions. Our eyes are sharp, although not as sharp as others. We can perceive a wide range of colors. More than some animals but not as many as others. This is not to say that our eyes are better or worse than others, it's simply that our eyes are evolved for what they need to do.

Dogs (and many other animals) have something we completely lack; a reflective layer in their eyes, behind the retina, called a *tapetum lucidum*. Figure 3 above shows how the tapetum lucidum is positioned. This membrane bounces light back towards the front of the eye, so photons that are not initially absorbed by cells in the retina on the first pass get a second chance to be absorbed. Photons that are not absorbed after bouncing off the tapetum lucidum are reflected out of the eye, and this is why dogs and other animals seem to have "glow in the dark" eyes if they are exposed to a beam of light, or a camera flash, at night. The reflective properties of the tapetum lucidum are achieved in different ways in different animals. The shark tapetum lucidum reflects light because it has guanine crystals in it, just like in the eyes of scallops and the scales of fish I previously mentioned. The tapetum lucidum of other animals is made of light-reflecting protein fibers, thin overlapping cells, or a variety of other kinds of crystals. No matter how it reflects light, it appears that the tapetum lucidum evolved independently in different animals and is another example of convergent evolution.

Dogs also have only two types of cone cells so they can see fewer colors than we can. They do not have red cones so they cannot detect red light and their vision tends to be "tinted" in the blue/yellow range compared to ours. But they're not color blind as the urban legend suggests. Moreover, dogs have more rods than we do, and since rods are arranged with multiple rods per neuron, dogs cannot see as much detail as humans can. On the other hand, the increased numbers of rods means that dogs have better night vision than humans.

The inability to detect red light is also found in some herd animals like white tail deer, which explains why hunters can get away with wearing blaze orange camouflage and still be "invisible" to deer when they are out hunting. Fortunately for the deer, they also have something that humans don't. Deer and many other animals (including dogs) have the capacity to see into the ultraviolet spectrum [62], which humans typically cannot. We cannot see UV light because the lens in the human eye physically blocks most of the light from the ultraviolet range of the spectrum from entering the eye [63]. The lenses in the eyes of deer and some other animals apparently don't have this UV blocking ability, and these animals can detect more UV light than humans can. Some hunters refuse to wash their hunting clothes in detergents that claim to "whiten" laundry since the chemicals used to whiten clothing often reflect UV light. If you have ever seen liquid laundry detergent glow under a black light you will know what I mean. If hunters use these chemicals on their clothes they will be like a shining beacon for deer or elk. The ability to see UV light probably evolved in animals that are more active at dawn and dusk, coinciding with the times when UV light is more prevalent.

Despite all this, a recent report shows that humans do, in fact, have an opsin protein that can absorb UV light [64]. So, it seems that humans might be able to see UV light if the lens of the eye didn't block UV light as previously mentioned. Interestingly, this is exactly what happens when a person has the lens of their eye removed for various reasons. For example, cataracts are a condition where the lens of the eye becomes cloudy so light cannot efficiently penetrate to the retina. Cataract surgery to replace the lens is very common these days, but in the early days of the technique, the lens was sometimes replaced with a material that does not block UV light. Individuals with this sort of replacement lens reported being able to see more "blue" colors, which was probably due to cones detecting UV light.

A famous case of this might have been in the early twentieth century when the painter Claude Monet was complaining that the flowers in his garden were no longer vibrantly colored, but "muddy." It turns out he had cataracts. In 1925, Monet had one of his lenses replaced in hopes that he would be able to resume painting the vivid colors of his garden. The famous water lilies he painted after the surgery are brighter than those he painted before the surgery and the petals are tinted in blue-white tones. It is believed that this color choice was made because the famous artist could now detect photons with wavelengths and energies from the UV spectrum [65].

No discussion of amazing eyes would be complete without mentioning the mantis shrimp. This little guy is remarkable for many reasons, notably its powerful punch that can actually break glass. But relevant to our discussion are its eyes. This may be one of the most interesting and simultaneously disappointing discoveries of all time. You might have heard that mantis shrimp can see far more colors than humans can. We have assumed this because mantis shrimp have somewhere between twelve and sixteen different opsin proteins, so in theory they could detect somewhere between twelve to sixteen different ranges of light waves. New evidence even suggests that some species of mantis shrimp might have up to thirty-three different opsins [66]!

The first letdown here is that mantis shrimp are not super special; we've recently identified other species that also have big numbers of opsins. Dragonflies and damselflies have between fifteen and thirty-three opsins, depending on species [67]. Water fleas have twenty-seven opsins [68].

Okay so they're not as special as we thought, but still, given what you now know about opsin proteins, you might be thinking to yourself, "wow, these animals must see with kaleidoscope vision!" With so many opsin proteins these animals must have the best eyes in the animal kingdom! Not necessarily. When mantis shrimp color perception was tested, they actually performed worse than humans. [69]. So, here's the next letdown, it appears that producing more opsins doesn't necessarily give an animal better color perception. It makes sense when you think about it; as long as you have the ability to perceive the entire rainbow of color with as few as three opsins, the others don't really add anything. True, mantis shrimp can detect UV light, but they don't appear to have the kaleidoscope of colors one might imagine. Next question then, what are all those opsins doing? The short answer is that nobody knows for sure, but it is speculated that perhaps all those opsins have something to do with allowing the mantis shrimp to control its super punch.

Opsins: An ancient gift

All of the opsins we have talked about so far function through the G-protein-coupled receptor (GPCR) cell signaling mechanism. There is, however, another other kind of opsin protein in bacteria and algae that acts like a channel protein, not a GPCR. In general, these are called *channel opsins*. Channel opsins use the same light-sensitive 11-cis retinal molecule and its "shape shifting" mechanism for activation, but when their retinal changes shape, it opens a pore in the opsin, much like the other channel proteins we have already looked at.

Channel opsins have structures that are very similar to GPCR opsins (Figure 8). Based on this similarity, we thought for a long time that GPCR opsins and channel opsins must be evolutionarily related. As scientists studied the DNA and protein sequences of more and more opsin and channel opsin proteins they found that, although the structures of the opsins are similar, the proteins themselves are built out of very different amino acid building blocks! It's as if two different builders built the same exact structure using different materials! Scientists decided that these two types of opsins were not actually related at all but were yet another example of convergent evolution. Over time our skills with comparing proteins have gotten even better, and the latest studies again predict that the two types of opsins might be evolutionarily related after all [70].

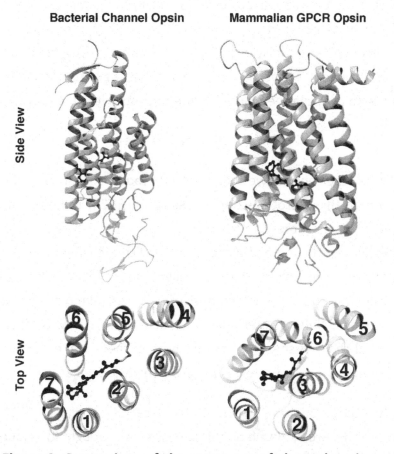

Figure 8. Comparison of the structures of channel opsin and GPCR opsin. Opsin proteins in bacterial cells and mammalian cells have different amino acid sequences, but similiar protein structures. Our best guess is these proteins are evolutionary conserved.

This back and forth of opinions is part of how science works, and it's a good thing. In the best-case scenarios, scientists may completely disagree about some piece of data until a novel experiment is performed that sheds new light on the topic. It doesn't always happen, but sometimes the disagreeing parties are actually (eventually) shown to have been both correct all along, each of them missing that last puzzle piece that allowed them to put their own results into context with their competitor's. In these cases, it's even better when the two labs publish their results side-by-side in the same journal. That is good science! Anyway, it seems like the opsins are all related, which is still incredible because it means that our vision is a gift from our ancient evolutionary friends, the bacteria.

Optogenetics: Mind control with light?!

In the early 2000's a few groups of scientists started really thinking outside the box and came up with a new way to use these channel opsin proteins for mind control, sort of. The field of *optogenetics* was born and it is one of the coolest bits of revolutionary science that I know of. Optogenetics is basically the science of planting artificial memories in animals using the power of light. Yes, you read that correctly. To help you understand how this works, let's do a thought experiment that sounds like something only Hollywood could dream up. Normal neurons like the ones in your brain do not have opsins and do not respond to light. Imagine taking a channel opsin protein from another creature and inserting it into a random neuron in your own brain. If you shine a bright light onto that neuron in your brain, will the channel cause the neuron to depolarize when light hits it?

The short answer is yes! It works because the new channel opsin protein depolarizes the neuron and fools the brain into thinking the neuron was stimulated by normal signals. Basically, the brain is confused because the normal stimulation and the artificial optogenetic stimulation both make the neuron fire in the same way. The brain can't tell the difference. This is what scientists are doing in the field of optogenetics. They are studying ways to control brain function by transplanting channel opsins into individual neurons of the brain so that they can poke and prod the brain with precision like never before. Sound like science fiction? It used to be, but not anymore. Its real, and this method is helping us to explore brain functions that have been impossible to study until now. A big problem in neuroscience is that it has always been difficult (or impossible) to specifically target one individual neuron in the brain. This makes it difficult (or impossible) to know exactly what you are doing when you poke around in the brain. It's like trying to fix the tiny gears in a watch with a pair of salad tongs. Until recently, our tools just lacked the precision to get the job done. Optogenetics is like a pair of little watchmaker forceps that let the watchmaker precisely place gears in exactly the right places.

I'm telling you about optogenetics because it's totally cool, but also to make the point that advanced technologies like this are only possible because we have a solid foundation of basic knowledge that came from a strong basic science research program. To make optogenetics work scientists needed to know about opsins in the first place, and this research has been going on for many decades. Scientists also needed to know about how DNA works in a cell and how to manipulate it; they needed to be able to precisely insert the DNA code for an opsin protein into individual neurons in a mouse brain. All this knowledge and know-how needs to come together for optogenetics to work and it is only possible by beginning with basic research, then being clever enough to imagine a new way to put all these parts together to make something completely new.

Regardless, now that we have optogenetics, what can we use it for? So far, optogenetics has allowed us to study some neurological diseases like epilepsy from a new direction. Optogenetics has also been used to help understand the neurological basis of drug addiction. Perhaps most impressively, this technique has helped us to understand the nature of learning and memory in the brain. Not to say we fully understand these things yet, but optogenetics gives us a new precision tool to dig around our consciousness with, and with new tools, usually comes new knowledge.

One of the truly amazing results of optogenetics studies so far is the successful implantation of a false memory in a mouse [71]. Scientists knew of a cluster of neurons in the mouse brain that was normally activated when a mouse received a small electric shock on its foot after being placed in a special electrified box. They hypothesized that if they could artificially activate those same neurons, they could teach the mouse to fear an electric box that it had never actually seen or been in. Scientists isolated the DNA that codes for a channel opsin protein in an algae cell and used genetic editing techniques to implant that channel opsin DNA into the mouse neurons that normally responded to the shock stimulus. Once they knew the new protein was successfully implanted in the neurons, they used a fiber optic cable to shine a light on the neurons, stimulating the channel opsins in the modified cells and causing the neurons to depolarize as if they were activated by a stimulus like an electrified box. When the scientists put the mouse into the box, it responded with fear because it "remembered" the false memory of getting shocked in there, even though it had never been in the box in the first place. The mouse learned to fear the box where it "received" the shock, even though the mouse was never in said box! Other mice that didn't have the opsin implant but were treated the exact same way did not have the same fear response.

Amazing right? In the classic science fiction movie *Inception*, the protagonists need to implant false memories into their target in order to manipulate the target's decision making in the future. What was once science fiction is now science fact. This first success is rudimentary, but future experiments will no doubt expand on the technology. More recently, scientists have "optogenetically" modified the nerves that control dragonfly flight. Now, the dragonflies can be controlled through little remote-controlled backpacks that emit light onto the nerves that control the dragonfly wings. Look it up online, it's a neat story. Who knows where the science will go next?

Could optogenetics be used in humans? Maybe in the future, but not anytime soon. So far, all the experiments have been performed in model systems where manipulation of the animals' DNA is easy, and non-controversial. Could it work in humans? Theoretically, yes. But in order to make neurons light sensitive, the channel opsin protein first needs to be inserted into the proper neuron(s). Right now we have two ways to do this. The most practical way is to use a genetically engineered virus to carry the opsin gene into the target cells. This is a little risky, since nearby "bystander" cells could easily get accidently dosed with the opsin DNA. If this happened, those cells would also become light sensitive and potentially lead to undesirable responses.

The second way is to introduce the opsin DNA into a person when they are nothing more than a one-celled fertilized egg. Obviously, this approach has ethical implications since it involves introducing the opsin gene before a person is even born. Moreover, this approach is no easy task. Nonetheless, it could be done. In 2018, Dr. He Jiankui, a Chinese scientist, claims to have modified the DNA in two young children for health reasons. Apparently, the children were born and are now living normal lives. If this is all true, these two people are the first genetically altered humans to be born. Stop and think about that for a second. The *"Brave New World"* envisioned by Aldous Huxley might be upon us. The work has not been published, again for ethical reasons, but the report seems to be genuine since Dr. Jiankui was sentenced to three years in prison for "illegal medical practice". We have been making transgenic animals for decades so it was only a matter of time before the world's first transgenic humans were born. That particular Pandora's box has been opened, for better or for worse.

In closing, lets recap the main points from this chapter. First, the "heart" of all visual systems that we know of is the conversion of 11-cis retinal to all-trans retinal when it absorbs a photon of light. This simple change in shape triggers the entire cascade by changing the shape of the opsin cage, and this is an excellent reminder of the basic premise in molecular biology; that shape dictates function. Second, you learned about the G-protein signaling mechanism, which is just the tip of the iceberg when it comes to cellular communication. Third, you learned about the molecular similarities and differences between our vision and that of other animals. Finally, you learned about optogenetics, the amazing new technology that takes our basic understanding of opsins and turns it into something entirely new that will help us learn more about the brain.

Finally, there is one more thing I'd like you to think about now, especially if you are new to molecular biology. Look at the list I just made for you about what you learned in this chapter alone. If you started reading with essentially no knowledge of molecular biology, this is an especially impressive bit of learning you have done! Of course, these topics go way, way, way deeper than I have gone here. But that doesn't matter. What's important is that you are beginning to understand the fundamentals of molecular biology! A friend once told me that his approach to teaching was to give students knowledge that "is like a river that is an inch deep and a mile wide". This means that his students should know a little about everything. Once you have that, you can go search various resources with more confidence for whatever information floats your boat. If you're new to this field, your river of molecular biology knowledge might be an inch deep but it's not a mile wide yet. Keep exploring to make your river wider.

Chapter 8: Taste and Smell

The corpse flower blooms only once every ten years or so. When it does, it emits the smell of dead meat to attract flies that then spread pollen to other corpse flowers.

Our senses are actually nothing more than a few receptor proteins that interact with the physical world and then send nervous signals to the brain. The senses of touch, hearing, and vision each have only a few types of receptors, and that works for those senses because vibrations, photons, and pressure are fairly "simple" stimulations that can all be sensed with only a few different receptors. In comparison, there are multitides of diverse molecules in our food and air that we need to be able to detect. To accomplish this, we need multitudes of different receptors and some interesting molecular biology.

Your taste isn't all that bad

You taste with your tongue, with the roof of your mouth, and maybe even with your uvula, that little flap of tissue hanging at the back of your mouth. Each of these regions in your mouth has *taste buds*, and these little organs are where your taste receptors are kept. To be accurate, those little bumps you see on your tongue are not actually your taste buds. The bumps are called *papillae* and they are a good way to introduce a really important concept: *surface area*. In biology, increasing the surface area of a structure increases the amount of exposure that structure has to its environment. In the lungs, a greater surface area means more room to interact with fresh air coming in. In the intestines a greater surface area means more opportunities to absorb nutrients passing through. In your mouth, a greater surface area means more chances for the molecules in the food you eat to interact with a taste bud. The papillae are there simply to increase the surface area of your tongue so more taste buds can be packed onto the tongue.

The taste buds themselves are found under the surface of the tongue, on the sides of pits within the papillae. Taste buds are composed of somewhere between fifty and one hundred individual cells. Each individual cell is topped with a cluster of hair-like structures called *microvilli* that look a lot like the stereocilia on hair cells. Microvilli are found throughout the animal kingdom and their job is almost always to increase the surface area of the cell to which they are attached. So, in addition to the pits within the papillae, the microvilli add even more surface area where molecules from the food you eat have a chance to interact with a receptor. The actual taste receptor proteins are located all along the microvilli. To put it simply:

papillae + microvilli = surface area = more taste receptors = more sensitive sense of taste = better eating experiences. As you eat, your food begins to break down and dissolve into the saliva. The saliva, carrying broken down molecules of food, leaks into all the nooks and crannies created by the papillae and microvilli, bathing those taste receptors in a soup of sugars, fats, and proteins from your food. It is these dissolved molecules that are the ligands of taste.

Each individual cell in the taste bud is attached to a different nerve fiber and, as we have seen multiple times now, receptor activation leads to depolarization of the cell, which sends a message back to the brain. Each taste cell makes only one type of taste receptor. Thus, each cell senses a single kind of molecule, or a single taste. But since a taste bud contains between fifty and one hundred individual taste cells, a single taste bud is usually able to respond to all the basic tastes. Taste buds are scattered about inside your mouth and across your tongue. This means that, contrary to popular belief, the tongue is not divided into

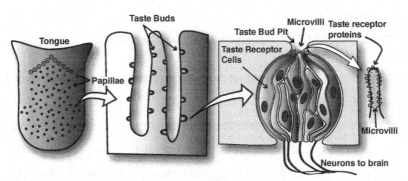

Figure 1. Anatomy of taste buds. *Taste buds are located in pits that go under the surface of the tongue. Taste receptor proteins are located on microvilli located at the top of taste receptor cell. Pits and microvilli work together to increase surface area and maximize our sense of taste.*

zones that each sense a particular taste. Rather, your ability to taste things is scattered around inside your mouth everywhere there are taste buds.

A receptor for every occasion

There are five basic tastes that humans can sense: sweet, sour, bitter, salty, and umami. Accordingly, receptor proteins have been discovered for sweet and umami [72], sour [73], and bitter [74]. The exact identity of a salt-specific receptor has eluded us so far. We also know that there are receptors for other non-taste qualities of food such as spiciness, coldness, and fat content. We already discussed two of these non-taste receptors; the TRPV1 channel (activated by heat, capsaicin, and resiniferatoxin) and the TRPM8 receptor (activated by menthol and cold). The incredible range of flavors and textures that we experience every day is achieved by simultaneously stimulating various receptors with various amounts of the basic taste molecules within the soup of our partially digested food. The blending of multiple inputs to create an average is basically the same mechanism used to perceive the color yellow even though we do not have a yellow cone protein.

Any given food will contain some, or maybe all of these taste molecules, in varying amounts. Since diverse foods have different amounts of taste molecules, they all taste distinctive. The result is that there seems to be an endless combination of tastes for us to experience. How many total flavors can you taste? It's not really known, but I'm sure that it's way more than the 31 offered at your local ice cream shop.

Oh how sweet it is: The sweetness receptor

Among the various taste receptors that we know of, we best understand the receptor for sweetness, or the *T1R2+T1R3 receptor* [75-77]. T1R2 and T1R3 are actually two separate proteins that stick together and make a larger protein called T1R2+T1R3. That large protein is the singular receptor for all the different molecules that our brain interprets as sweet. This is really interesting. I have been telling you all along that receptors usually have a single ligand that they can bind to. The sweet receptor is an exception. As it turns out, a very wide variety of molecules that we consider to be "sweet" bind to the T1R2+T1R3 receptor. Natural sugars, like glucose, sucrose, and fructose, plus many artificial sweeteners, like sucralose, saccharine, and aspartame, all bind to this receptor. Hopefully, you are asking yourself the question, "how can all these different molecules bind to the receptor's active site if they all have different chemistries and shapes"? Good question.

The answer is that different sweeteners bind to **different places** on the **same receptor**. T1R2+T1R3 is different from the majority of receptors since it has **several active sites** to bind its many ligands [78].

The splendid molecular biology of Splenda

Even though the T1R2+T1R3 receptor has multiple active sites, many different ligands can also use the same active site. For example, you have likely heard of *sucrose* (common white sugar), *glucose*, and maybe the artificial sweetener *sucralose* (found in Splenda). While all of these molecules are "sweet", they do not all have equal sweetness. Sucrose is the industry standard by which the sweetness of all other sugars is based, and we therefore define the sweetness of sucrose as 1.0. Glucose is less sweet than sucrose with a sweetness of 0.75. In comparison, sucralose has a sweetness of 600! From a practical perspective, if you used pure sucralose to bake cookies you would need hardly any at all. Substituting sucralose directly for sucrose would dramatically change the physical bulk and consistency of your cookies. To compensate for this, granulated Splenda contains both sucralose and another chemical called *maltodextrose*. The maltodextrose doesn't taste like anything. It's just added to dilute the sucralose so that it can be substituted for sucrose in baking without having to change the amount in the recipe.

Even though sucrose, glucose, and sucralose have different shapes, they all bind into the same active site on the same receptor, and this raises an interesting question: how can they have such different sweetnesses if they bind to the same site? The answer is in how **tightly** they bind to the sweetness receptor. Sucrose binds about two and one-half times more strongly than glucose and correspondingly, sucrose is sweeter than glucose. Sucralose binds about eight times more strongly than glucose, and, is correspondingly 600 times sweeter than glucose [79]. The reason for this is that if a molecule binds to its receptor very tightly, it stays attached longer and can activate the receptor more effectively. It can send a bigger, stronger signal to the brain. In the case of sucralose, a ligand-receptor bond that is eight times stronger adds up to a massive increase in receptor activation.

The ability to bind tightly to a receptor is a really important consideration in molecular biology and especially when scientists are making new drugs. In general, the more tightly a ligand binds to its receptor, the more specific it will be for that receptor, and the less likely it will be to accidently bind to other receptors and cause unintentional side effects. Ligands that bind to their receptors more tightly can also be used at lower concentrations. If you considered sucrose and sucralose to be "drugs", sucralose would be a much more effective and specific drug than sucrose.

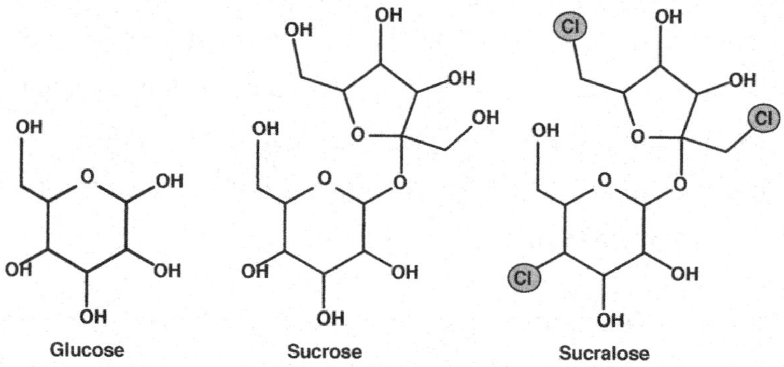

Figure 2. Structures of glucose, sucrose, and sucralose.

Even though sucralose is so much sweeter than sucrose, the chemical structures are quite similar. Sucralose is actually a lab-modified version of sucrose that has three chlorine atoms attached to it. These chlorine atoms have a double benefit; they increase the affinity of sucralose for the sweetness receptor (and thereby make it sweeter) and simultaneously change the shape of the molecule such that it no longer fits into the active site of *sucrase,* the enzyme that normally breaks down sucrose. Because of this modification, sucralose cannot be digested by your body, and therefore it contains zero calories.

The miracle sweetener

Speaking of zero-calorie sweeteners, there is one more that I need to tell you about. This one is a real miracle. The sweetener is called *miraculin* and it comes from a small red berry called a *miracle berry*. When you eat a miracle berry, or the miraculin protein, you won't notice much at first. But if you drink or eat something sour that has acid in it within the next hour you'll get a surprise: you'll taste sweet instead of sour. Is it a miracle? Not really. Again, it's just molecular biology.

Miraculin binds to the same T1R2-T1R3 sweetness receptor we've been discussing [80]. When it binds, it cannot quite access the active site in a way that would activate the receptor. However, if some acid, like the citric acid in lemon juice, is introduced, miraculin changes shape and inserts part of itself into one of the T1R2-T1R3 active sites, triggering a sensation of sweetness instead of sour.

But wait, it gets better! If you were to wash the acid from your mouth and try to eat some real sweetener, like glucose or sucrose, the sweetness would be blunted [80].

But wait again, it gets even better! Scientists have now created *transgenic plants* by isolating the gene that codes for the miraculin protein and splicing that gene into the DNA of tomato and lettuce plants [81, 82]. The result, of course, is sweet tomatoes and lettuce that children might actually like to eat but is still good for them! A miracle indeed!

This is a good time to ask you a question to see how you are doing with your molecular biology. How can miraculin be sweet when it's in an acidic environment but also anti-sweet when it's not in acid? Give me some SWAG! This is the best kind of question because, as far as I can tell, scientists don't know yet.

My SWAG is based on what I have already told you about the T1R2-T1R3 receptor. Remember that this receptor is unusual in that it seems to have more than one active site where sweet molecules can bind to and activate it? Based on this, I *hypothesize* that miraculin binds to the T1R2-T1R3 at **two different positions**. At one position, when there is no acid present, it blocks one active site where some sweet molecules usually bind. Here it is anti-sweet; it physically blocks other "sweet" molecules from reaching the receptor. At the other position, and when acid is present, it changes shape and inserts itself into the active site to turn the receptor on. Here it becomes sweet. Can you think of any other possible mechanisms?

There is one more interesting thing to say about miraculin. This molecule was described hundreds of years ago and officially scientifically characterized in 1968 [83]. For a miracle molecule that is all natural, could potentially change the way we sweeten our food forever, and could help solve the obesity and type 2 diabetes epidemics in the United States, why haven't most people heard of it?

It seems that there is something of a conspiracy theory floating around about miraculin. I'll let you research it for yourself [84], but the upshot is that the founders of Miralin, a company that was set to commercialize miraculin as an artificial sweetener, believe they were sabotaged by the sugar industry. If true, this wouldn't be the first time the sugar industry was caught with its hand in the cookie jar. Another episode of misdeeds by the sugar industry that has more concrete evidence happened in the 1950s. According to documents discovered in 2016 [85], the sugar industry paid three Harvard scientists to write an article that described the evils of fats in the diet [86]. This article helped usher in the low fat/high sugar diet that was popular in the United States in the 1980s and 1990s but has now been shown to have serious health consequences. This story ran through the media in 2016, and the New York Times has a nice article [87] that summarizes the conspiracy nicely.

So, what's the point of telling you about the transgressions of the sugar industry? Scientists such as those who took the sweet money of the sugar industry to author their infamous paper are betrayers of the public trust. Historically, scientists are some of the most trusted people in the world. According to a 2020 poll done by the PEW Research Center, medical scientists (think doctors) and scientists (think professors) are the first and second most trusted professions in the United States [88]. For various reasons, these days it seems like mistrust is a regular state of being. We need people like scientists who maintain their trustworthiness and are reliable purveyors of information. Scientists like those that sold out to the sugar industry do a great discredit to this trust. Thankfully, the vast majority of doctors and scientists are committed to keeping the public trust. They might still get things wrong occasionally, but more often than not either the scientists will notice and try to correct their errors or, eventually, the scientific method will uncover the error. Nonetheless, it is always wise to make sure that a scientific claim is supported by data from multiple independent labs before putting too much faith in the findings.

Other taste receptors

I have now described in some detail the molecular biology of your sense of sweetness, but what about the bitter, sour, salty, and umami tastes? Receptors for each of these flavors (except salty) have been identified, although little is currently known about the receptors for sour and bitter. Nonetheless, there are a few things that I think you will find interesting and will provide good opportunities to learn more about molecular biology.

Yum yum umami

The umami receptor was identified in 2002 by two independent groups [89, 90]. This was a nice example of two competitive groups that confirmed each other's results by finding the same receptor protein. When multiple publications independently verify a finding, you can be more confident that the results are legitimate.

Like the sweet receptor, this umami receptor is made from two different proteins that stick together in the cell membrane. The umami receptor is called *T1R1+T1R3*. Don't be confused since this name is very similar to the name for the sweet receptor (T1R2+T1R3). Names like these can be daunting for non-scientists since they don't have an obvious meaning. But the meaning is there if you take a closer look. They both have the "T1R3" part in their names for a good reason. Both the umami and sweet receptors contain the same T1R3 subunit in their overall structure. What makes them different is the other part.

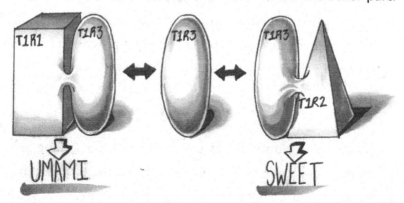

Figure 3. Mixing and matching of the umami and sweetness receptors.

The T1R1 protein is in the umami receptor, and the T1R2 protein is in the sweet receptor, and both are independently paired with T1R3. This sort of mixing and matching is not uncommon in molecular biology. Molecules often have multiple binding partners that produce different outcomes.

A bitter pill - bitter receptors

While we have one receptor each for sweet and umami flavors, it seems like we have somewhere between forty and eighty different receptors for bitter flavors! It is speculated that this wide variety of receptors is an evolutionary adaptation to make sure that we can taste poisonous molecules since poisons are often bitter.

Interestingly, these bitter receptors are G-protein-coupled receptors, just like the opsin proteins we encountered when we discussed vision. In fact, I didn't mention it before, but the umami receptor, sweet receptor, and many of the receptors that we use for smelling are also GPCRs. There are an estimated eight hundred different GPCRs in the human genome! More than any other type of receptor protein. Perhaps it is because of this diversity and importance that the majority of our pharmaceuticals work by changing the activity of GPCR receptors.

What your nose knows

As flexible and wonderful as our sense of taste is, our sense of smell is far more versatile. It's estimated that humans have about four hundred different types of receptors in our noses for detecting different types of smell molecules in the air [91]. Most of the things we smell everyday contain multiple odor molecules dissolved into the surrounding air. What is your favorite smell? My favorite might be freshly baked bread. Scientists have analyzed the odor molecules that come out of a loaf of fresh bread and have found dozens of different molecules. These dozens of molecules interact with perhaps dozens of different receptors and your brain interprets the particular combination of nerve depolarization as "fresh bread". Considering that humans have about four hundred different odor receptors, imagine how many combinations are possible!

Given all these combinations, it's estimated that humans can literally perceive trillions of different scents [92]. That's amazing, but our sniffers are far from the best. It's estimated that a bloodhound has approximately eleven hundred individual kinds of smell receptors, and forty times more total smell receptors than humans. No wonder these dogs are commonly referred to as "a nose with a dog attached".

Our smell receptors are found in a little patch of tissue called the *olfactory epithelium,* which is found at the top of your nasal cavity. The cells that express smell receptors in this tissue are really unique. They are called *olfactory sensory neurons* and they are topped with many long, tentacle-like appendages called *cilia*. Similar to the microvilli that sit atop the taste cells in taste buds, the function of cilia is to make our sense of smell more sensitive by increasing the surface area, giving odor molecules more chances to find a receptor.

Shown on the next page are some images from the nasal cavity of an alligator. Believe it or not, the cells of your olfactory epithelium are very similar to the sensory cells in an alligator's olfactory epithelium. In between the olfactory sensory neurons are other cells called *support cells* that, as their name implies, provide physical and metabolic support for the olfactory sensory neurons. An interesting note about these support cells is that infection of these cells by the SARS-COV-2 virus seems to be the reason people sometimes lose their sense of smell as a symptom of Covid-19. These support cells express two proteins, *ACE2* and *TMPRSS2*, on their surface that the virus uses as a receptor to bind to the cell surface [93, 94]. When the virus gets in and kills the support cells, the olfactory sensory neurons stop working. Luckily, after the viral infection is over, support cells grow back, and most people regain their sense of smell.

Just as in our senses of taste and vision, each cell in the olfactory epithelium expresses only one type of smell receptor protein. Each of these smell receptor proteins is activated by a single, or maybe a few, chemically similar odor molecule(s) that are

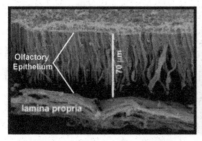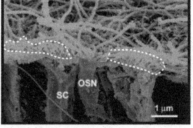

Figure 4. Images of alligator olfactory epithelium. The image on the left shows a low magnification view of the olfactory epithelium standing on the basement membrane (lamina propria). At 70 micrometers (um), these cells are quite long compared to other cells. The image on the right shows a higher magnification of the olfactory epithelium that has both olfactory sensory neurons (OSN) and support cells (SC). Notice the tentacle-like cilia on top of the sensory neuron and the much smaller microvilli (dashed area) on top of the support cell. Open access image taken from [95] with slight modification.

dissolved in the air. In other words, just like the TRPV1 and $Na_v1.7$ channels, these smell receptors are also activated when a physical ligand binds to them. As I mentioned before, the majority of the known smell receptors are G-protein-coupled receptors. Like the rods and cones, each individual olfactory sensory cell is directly connected to the brain [96], so the overall scent that we perceive is an **average** of the many different odor molecules that are coming off of the thing we are smelling. If the sensory cells in the olfactory epithelium each expressed more than one kind of receptor protein, we would not be able to discriminate between different smells as accurately. Everything would smell like a mixture of multiple scents. It would be sort of like when you mix all the colors on your paint palette together and you get an average of all the colors, which is usually brown.

Despite this, this is actually how we smell! Odors are very complex and consist of many individual molecules that sort of make an average "smell impression" in your brain. This is because most things we smell are composed of many different odor molecules and your brain has to simultaneously integrate input from maybe dozens of smell receptors at once. In addition, like taste receptors, a single receptor in your nose can sometimes be activated by several different odor molecules. What's more, a singular odor molecule can sometimes bind to several different receptors, triggering multiple perceptions at once. As in the case of taste receptors, olfactory sensation is an exception to the rule of "one-receptor-one-ligand". Even though each cell in the olfactory epithelium only makes one kind of receptor, there is mixing and matching happening between different odor molecules and different receptors. The result is the incredible diversity of smells that we experience every day.

Molecules of smell

For each of the receptors we have discussed, we have examined some of their natural ligands. Let's keep this going and talk about some of the molecules that activate our smell receptors. In general, odor molecules are small molecules that readily evaporate into the air. Perhaps the most famous odor molecule is *isoamyl acetate*. This molecule produces the quintessential banana smell. It's a molecule that some yeasts produce, giving those Bavarian-style wheat beers their characteristic banana smell, and any student who has taken organic chemistry has probably synthesized it in their lab.

If you make just a few chemical modifications to isoamyl acetate, you get isovaleric acid. Bacterial cells on your skin metabolize molecules in your sweat to produce isovaleric acid which is primarily responsible for the smell of stinky socks. Interestingly, isovaleric acid is also found in many of the worlds "best cheeses" where it is prized for the strong smell it imparts to some cheeses.

On the other hand, *cadaverine* and *putrescine,* which smell like rotting flesh and are produced by bacterial cells as they break down meat, also have similar structures. But unlike isoamyl acetate and isovaleric acid, these two molecules smell the same, which makes sense since they both interact with the same receptor protein [97].

Although it seems like they should be in there, neither cadaverine nor putrescine have been found in the infamous corpse flower. The corpse flower is the largest flower in the world and only blooms every seven to ten years. When it does, it produces the smell of rotting flesh to attract flies to carry its pollen to other flowers. Scientists analyzing the molecules in the corpse flower scent did not find evidence of cadaverine or putrescine. Instead, the main odor-causing molecule was identified as *dimethyl trisulfide* [98], which is also thought to be the primary odor molecule that blow flies use to locate corpses in which they can lay their eggs [99]. Interestingly, dimethyl trisulfide is also produced by some forms of cancer, and dogs can be trained to smell dimethyl trisulfide in a person's breath to detect some forms of lung cancer [100, 101].

Finally, *skatole* is the molecule primarily responsible for the smell of feces. But that's only when you get a big whiff of skatole. Strangely enough, if you get a small dose, it can smell

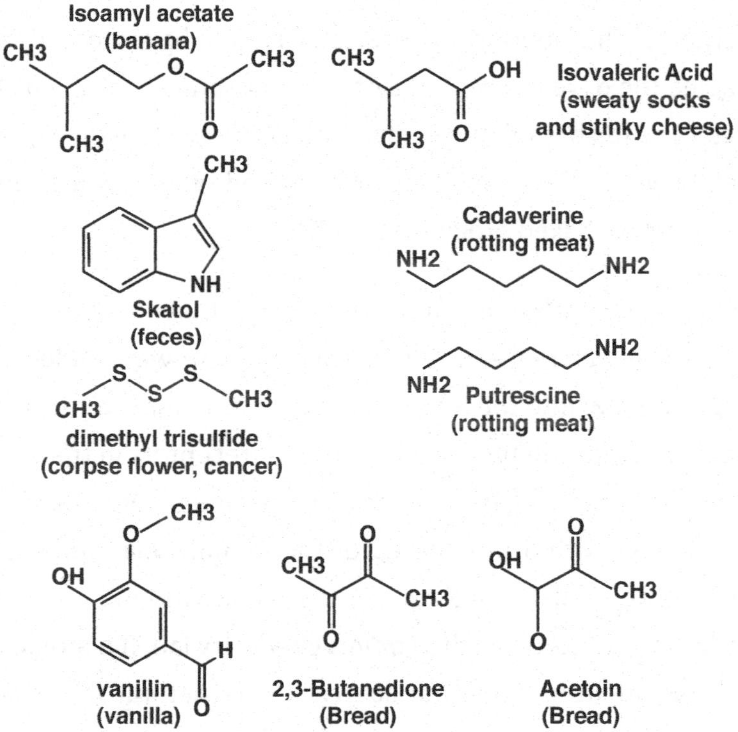

Figure 5. Some of the more interesting molecules we can smell.

more like a flower. Very low levels of skatole are actually used as an additive in some ice creams, perfumes, and cigarettes to give them that little something extra! Skatole is also known to attract flies to poop and some flowers use skatole for "*dung mimicry*" in order to attract flies to their flowers to help spread their pollen.

Obviously, scientists have a strong interest in discovering and making odor molecules for various purposes. Because of the powerful (and visceral) perceptions associated with molecules like putrescine, cadaverine, and skatole, these chemicals are commonly added to "stink bombs," and various militaries are rumored to be making military-grade stink bombs with these powerful odorants as the primary weapons. On the other hand, analysis of the odorant molecules in freshly baked bread has enabled the development of air freshener sprays that smell like "freshly baked bread" and contain *vanillin, 2-3 Butanedione*, and *acetoin* which are some of the buttery and sweet components of the freshly baked bread smell [102].

On a final note, it's important to remember that no matter how a molecule was made, you sense it the same way. Molecules simply dissolve into the air, get into your nose, and interact with one or a few of the hundreds of odorant receptors in the nasal cavity. Or they dissolve into your saliva, get into your taste buds, and interact with a taste receptor there. Individual molecules smell, taste, and behave exactly the same whether they are produced in nature's kitchen or in a chemistry lab. If a molecule is extracted from a natural source it's a "natural flavoring", whereas if it's synthesized and purified by humans it's called an "artificial flavoring". In reality, these are all just molecules that bind to a receptor and result in the perception of a smell or taste, no matter where they come from. Despite the simplicity of this setup (molecule + receptor = perception), natural smells usually consist of many odorant molecules that combine to produce complex smells that cannot easily be fully recreated with a single or even a few artificially created molecules. You will never fool me into eating a lab synthesized loaf of "fresh bread".

Chapter 9: "Superpowers"

Whales and many other animals can detect some energies and stimuli that humans cannot. Very low frequency sounds of some submarine sonars may be loud enough to disorient whales and cause them to beach themselves.

W e are bombarded by information every millisecond of every day. In the previous chapters we discussed how we perceive the information that comes in the form of sound waves, photons, pressure, or various molecules in our air and food. But are there other kinds of information in the world that we cannot detect?

There are many stimulations and energies in the world that humans cannot detect. Remember that the section of the electromagnetic spectrum that we can actually see is only a small range of the entire electromagnetic spectrum. As a matter of fact, at this very moment, very short wavelength radiations, like gamma rays, and long wavelength radiations, like radio waves are passing through your body and you are totally unaware of them. Infrared radiation is being emitted by every surface you can see, but you can't detect it. And when you watch your food in the microwave, you don't see beams of microwaves zapping your food. Even if your face is smooshed against the glass so you can see your pizza rolls get melty.

We can't see these things simply because we lack receptors that can interact with these energies. When it comes to these invisible energies, we are like blind people living in the world with no visual sense of the world around us. We can hear, feel, etc. but not see what is right in front of our eyes.

Of course, not everything we detect is part of the electromagnetic spectrum. Piezo proteins sense physical distortion, and smell and taste receptors detect molecules in your air and food. Is it possible there are other energies or stimulations we cannot even imagine? What about ghosts and other paranormal activities? Are ghosts actually some undiscovered form of energy that humans cannot detect because we lack a receptor for it? What about extra sensory perception (ESP)? Is it possible that some people are actually sensitive to an energy that most of us cannot detect? These questions have entertained people for many years, but so far scientific approaches and investigation have debunked every claim of the existence of supernatural energies. Nonetheless, most scientists will tell you that they would be happy to accept evidence for the supernatural if it was presented in a scientific and reproducible way and was able to be explained. After all, the word "supernatural" literally means things that are beyond the natural, right? If the "hidden energies" in ghosts and ESP could be detected with receptor proteins, they would only be "natural".

Some animals are able to detect these hidden energies that humans can only imagine. For instance, many animals detect and use infrasonic sounds that humans cannot detect. Infrasonic sounds are sounds with frequencies below the lower limit of human detection which is about 20 Hz. The best understood examples of infrasonic sounds are communications between elephants, which can travel for many miles and are often sensed by the elephants feet through vibrations of the ground. One of my favorite examples of infrasound are the sounds that a roughed grouse makes when it beats its wings on a "drumming log". You can "hear" these sounds in the fall of the year when male grouse are searching for a mate. I say "hear" but when you experience this, you feel the vibrations in your chest more than you really hear them in your ears. It's quite remarkable. Natural events like earthquakes, thunder, and avalanches also produce infrasonic sounds and this is probably the reason why many animals can sense earthquakes even before we can feel the earthquake. Despite knowing that some animals can detect infrasound, it is not yet known what cellular and/or molecular mechanisms are responsible.

Given the sensitivity of many animals to infrasonic sounds, our modern world of technology can sometimes interfere with everyday animal behaviors. For example, some sonars used in submarines, seismic surveys of the ocean floor, and boat propellers all produce significant infrasonic sound [103]. These very low frequency sounds can be heard by whales hundreds of miles away. Even at this distance, the sounds can be around 140 decibels. For comparison, chainsaws run between 100-120 decibels. Rock concerts reach about 130 decibels. Some think that these sounds scare whales or damage their navigation senses and have been attributed to causing the animals to beach themselves [104]. Other examples of manmade infrasonic sounds come from wind turbines, cars and trucks, and nuclear explosions.

Infrasonic sounds are not the only examples of how our modern world accidently impacts animals. Other examples include light pollution from our cars, cities, and houses and low-level water contamination from various chemicals that are not removed from our water treatment facilities. For a longer discussion on this important topic, check out this reference [105].

Clearly, animals are more sensitive to some things than we are. How do they do it? They just have receptor proteins that we lack. We have already discussed how skates can sense electric fields with the $Ca_v1.3$ receptor, how bats and dolphins can detect ultrasonic vibration with prestin, and how and why many animals can see UV light. The goal of this last chapter is to shed some light on other animal superpowers by examining their molecular biology. We don't yet fully understand the last two senses we are discussing in this chapter, infrared detection and magnetic field detection, so I'm going to run you through what we do know. From there you might have to make some hypotheses of your own.

Seeing in the "dark"

The first "superpower" is the ability of some animals to detect infrared radiation. This basically means these animals are able to "see" heat. Humans can do this too, but only with the help of infrared cameras. I bet this ability would get even more exciting if we fully understood how it works! Unfortunately, we are only beginning to learn about the molecular mechanism of infrared detection.

Infrared radiation gets its name because the wavelength of the radiation is just a little bit longer than red light. We don't call infrared radiation "light" simply because we can't see it with our eyes. Our opsins simply cannot capture photons in that part of the electromagnetic spectrum. Nonetheless, it's basically the same as red light, but with a longer wavelength.

It was originally thought that the ability to see infrared radiation would involve some form of retinal-opsin system, similar to the system our eyes use to see radiation in the visible spectrum. In 2010, researchers wanted to discover the receptor that could detect infrared light. They started by finding a good model system, which in their case involved snakes. Some snakes are called "pit vipers" because they have two organs on the front of their faces, usually between their eyes and nostrils, called *pit organs.* It was discovered that pit organs were sensitive to heat way back in 1937 [106], so this seemed to be a good model in which scientists could discover an infrared receptor. The researchers compared which genes were turned on in the heat-sensitive tissues compared to the genes turned on in similar but non heat-sensitive tissues. In one of the rarest of results, the kind that most scientists salivate for, they found only one single gene (of all the genes in that snake!) that was "on" in the heat-sensitive tissues and "off" in the heat-insensitive tissues. That gene codes for the snake *TRPA1* receptor protein [107]. If you remember, we talked way back in chapter 2 about TRPV1, which is the receptor protein for capsaicin and heat, and TRPM8 which is the receptor for menthol and cold. TRPA1 is related to TRPV1 and TRPM8. Human TRPA1 is known as the "wasabi receptor" [108]. It has been identified as the receptor for *allyl isothiocyanate*, the stuff that makes our noses burn when we eat wasabi, mustard, or horseradish. Interestingly, years after its initial discovery, TRPA1 was also recognized as one of several different cold receptors since it was shown to be activated by extreme cold temperatures [109]. Regardless of where it was discovered, when the researchers transplanted a copy of the snake version of the TRPA1 gene into cells and exposed them to infrared light, lo and behold those cells were now sensitive to infrared light [107]. Since the first predictions were that the infrared receptor would be an opsin protein, this must have come as a surprise.

To try and figure out why and how the pit viper TRPA1 protein is sensitive to heat, while the human TRPA1 is not, scientists used some genetic engineering tricks and swapped sections of the pit viper TRPA1 protein with the human TRPA1 protein. They found a piece of the pit viper TRPA1 protein that, when attached to the human TRPA1 protein, made the human protein sensitive to heat [110]. Exactly how this little piece of the pit viper TRPA1 protein actually senses infrared radiation is still not known. A good guess however would be that the heat associated with infrared radiation somehow changes the shape of the protein causing it to open and trigger cell depolarization. Regardless of how it senses heat, this example illustrates an important lesson about how most proteins are built. Proteins like the TRPA1 protein are often built out of connected subunits that have their own activities. These individual activities all cooperate to give the protein its final function. These activities are sometimes maintained even when the subunit is transplanted onto another similar but different molecule.

Since finding that TRPA1 was responsible for detecting heat in pit organs, researchers have been looking to see if TRPA1 might also be used to detect heat in other animals. So far they found that mosquitos also use TRPA1 to help them find food [111]. These little monsters first use G-protein coupled receptors to sense airborne carbon dioxide and octenol emissions (a common molecule in sweat) from animals. Carbon dioxide and octenol emissions make a chemical trail in the air that the mosquitos follow to get close to their dinner [112]. Interestingly, some commercially available mosquito traps actually use octenol as bait to lure mosquitos to their death. Once the mosquitos are close, they use their TRPA1 channels to determine if the potential meal is not too hot and not too cold, just like Goldilocks. By the way, the common mosquito repellent DEET seems to work by either decreasing the amount of octenol that evaporates from your skin and/or binding to a different mosquito receptor protein (not TRPA1) that activates an avoidance behavior in mosquitos, sort of like mosquito pepper spray [113]. Scientists are not sure yet how it works, but they do know that DEET does not block the TRPA1 receptor from interacting with octenol. Also, don't confuse octanol (with an A) which was discussed in chapter 6 and smells like roses and oranges with octenol (with an E) which smells like sweat.

This protein still has many things to teach us about molecular biology. TRPA1 is somewhat of an anomaly among receptor proteins since so many different things activate it. I have already mentioned that it is activated by allyl isothiocyanate, cold, and heat. Allyl isothiocyanate, is an irritant that is used by plants to prevent pests from eating those plants that contain it. Think about the negative effects of cutting onions, that's allyl isothiocyanate. It's also one of the primary ingredients in tear gas! In addition to allyl isothiocyanate, TRPA1 is also a receptor for many different kinds of irritants. For instance, sodium hypochlorite (bleach), cinnamaldehyde (cinnamon), hydrogen peroxide, and acrolein (wood, cigarette, fat smoke) have all been shown to activate TRPA1. In the figure on the next page, I have drawn some of these molecules so you can appreciate that at first glance their structures appear to have almost nothing in common. This raises the question, how can TRPA1 be activated by these different shaped molecules?

If you remembered what I taught you about taste receptors and guessed that they have multiple binding sites for a variety of molecules, that would be a good guess. But you would be wrong. The actual answer is that these chemicals interact with TRPA1 based primarily on their chemistry, not so much based on their shape. If you look closer at figure 1, notice that all the structures have either sulphur or oxygen atoms. Sulphur and oxygen pull electrons away from carbon and hydrogen atoms and make these atoms want to take electrons from other places. Molecules like these are highly reactive. They are *electrophiles*, molecules desperately searching for electrons to form bonds with, and TRPA1 has a very special electron donor "bait" in it's active site that attracts these electrophiles like flies to skatole!

The ligand binding site of TRPA1 is set up in such a way that there

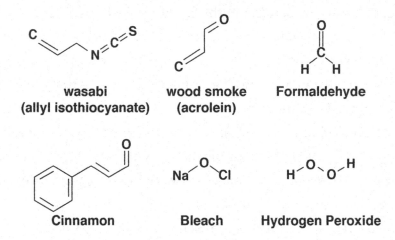

wasabi | **wood smoke** | **Formaldehyde**
(allyl isothiocyanate) | **(acrolein)**

Cinnamon | **Bleach** | **Hydrogen Peroxide**

Figure 1: Some irritant molecules of smell that interact with TRPA1.

is one particular amino acid called cysteine 621 that is extremely reactive with electrophiles. The chemical environment of this cysteine residue inside the protein makes this site about six thousand times more reactive than it would otherwise be [114]. When an electrophilic irritant gets anywhere close, it is extremely likely to form a new chemical bond with TRPA1. Ligands that bind based on electrophilicity, not shape, are somewhat different than the "key in the lock" or the "last puzzle piece" models of an active site that we talked about earlier. In this case, the "key" can be pretty much any shape, as long as it has the right chemistry and is roughly the right size.

This unique mechanism for ligand selection has some very important consequences. You see, TRPA1 is present in your eyes and airways. Its ability to react with a wide variety of irritants makes TRPA1 very good at constantly surveying the air you breathe for irritants. Think about the last time you were sitting around a campfire and smoke blew into your face. The coughing and burning eyes you experienced were probably caused, at least in part, by *acrolein* in the wood smoke binding to TRPA1 and causing an inflammatory response. As much as smoke in your eyes is annoying, this is actually really exciting for people who study asthma. Doctors have known for decades that asthma is triggered by various molecules in smog, air pollution, smoke, and even cold air, but they have not understood the mechanism through which all these individual molecules affect asthmatics. With this new understanding of how TRPA1 recognizes diverse electrophilic irritants such as acrolein, we may finally have uncovered the connection between air pollution, cold air, and asthma [115]. As a matter of fact, scientists recently deleted the TRPA1 protein from lab rats and discovered that TRPA1 is required for the rats to respond to an experimental model of asthma [116]. Stated simply, hypersensitivity of TRPA1 might be the elusive root cause of asthma. If findings like these continue, it won't be long before reseachers begin searching for molecules to block TRPA1 function and perhaps finally treat asthma at its source rather than just its symptoms.

The TRPA1 receptor is notable for a few more reasons beyond its role in infrared and irritant sensing. As you might imagine, any receptor that can cause you to feel like you got a big whiff of wasabi up your nose is a good candidate for toxin development by Mother Nature. This is exactly the case for both a recently discovered scorpion toxin that activates TRPA1 [117] and a tarantula-derived toxin that actually blocks TRPA1 activation [118]. The tarantula TRPA1 blocking toxin might be a good candidate for new pain medications or maybe doctors will someday be treating asthma with a new class of drugs based on tarantula toxins? Finally, if you ever get a chance to eat Szechuan buttons, which are some of the strangest things you will ever eat, you should know that the active ingredient, *spilanthol*, is also making your mouth feel funny because of its interactions with TRPA1 and TRPV1 receptors in your mouth [119].

Magnetoreception

Finally, we come to what may be the final frontier of sensory perception: magnetoreception. Magnetoreception is the ability some animals have to sense Earth's magnetic field. It is believed that these animals use Earth's magnetic field as a compass to enable them to navigate over long distances. For example, migratory birds are thought to use Earth's magnetic field to navigate on their yearly migrations. Turtles, salmon, and trout are also thought to use Earth's magnetic field to return to the same beach or river where they themselves were born, to lay their own eggs.

From the scientist's perspective, these observations are not proof of magnetoreception, just correlation. Perhaps these animals are navigating not based on the magnetic field at all, but by visual landmarks, or odors? For a long time, some people even argued that animals couldn't possibly sense Earth's magnetic field because the field is too weak and there is no way for a biological system to be sensitive enough to detect it. In 1965, however, the first experimental evidence for magnetic reception in birds started to appear [120]. Since then, many other animals including Monarch butterflies, salmon, lobsters, bats, and mole rats have been tested and have been proven to have the ability to sense Earth's magnetic field. Most recently, some sophisticated work has been trying to determine if even humans can detect magnetic fields with some sort of vestigial magnetic sensory system that we are not aware of. For the record, the newest evidence suggests that yes, our brains can respond to magnetic fields even if we are not aware of it [121]. Today, it is pretty well accepted that many animals can sense Earth's magnetic field, but many questions about how this works on the molecular level remain.

Since it is not yet settled how the magnetoreception works, this seems like a good way to close this book. It will give us an opportunity to think about what a magnetoreceptor sensory system might look like.

Think about all the sensory systems we have discussed so far. Without exception, all those systems involve a receptor protein that is able to interact with an external stimulus, cause a cell to depolarize, and send a message back to the brain. More than likely, the magnetoreceptor sense is going to be similar. The key piece of knowledge that is missing right now is the identity of the receptor, or receptors. I say receptors because, as we have seen, convergent evolution sometimes has a way of popping up where it is least expected. In this case, magnetoreception might very well be yet another example of convergent evolution and there may be multiple versions of such a receptor.

There are two major competing hypotheses for how a magnetoreceptor protein might work. The first is perhaps the most obvious and involves tiny crystals of *magnetite* (naturally occurring magnetic iron) attached to receptor proteins inside magnet-sensitive cells. Magnetic particles like these have been found inside of specialized bacterial organelles called *magnetosomes,* which enable some bacterial cells to sense Earth's magnetic field.

Since this is known to happen in bacterial cells and given the evolutionary relationships between bacterial cells and animals,

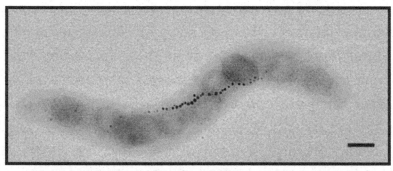

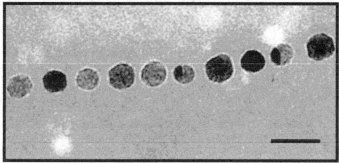

Figure 2: Bacterial Magnetosomes. *Magnetosomes such as these allow some bacterial cells to orientate themselves in the Earth's magnetic field. Images taken from [122].*

it seems logical that magnetoreception in animals might also involve magnetic iron particles. The simplest hypothesis is that these little magnetic particles might be attached to a receptor protein and cause the receptor to open or close in response to the Earth's magnetic field. In support of this, crystals of magnetite have been identified in the nose of rainbow trout [123] and these crystals do seem to respond to magnetic fields, although the fields tested were almost 5000 times stronger than Earth's [124], so it is currently unknown if these structures respond to Earth's field. Nonetheless, this paper did not identify a specific protein that could be activated by magnetic fields.

Magnetoreceptors that contain iron are the most straightforward hypothesis for how a magnetic field receptor might be able to physically sense magnetic fields. This idea makes a lot of sense, and it is intuitively satisfying since we are all familiar with magnets, but it's not the only option. There is another way that scientists believe could be used to sense magnetic fields. The second hypothesis involves a protein called *Cry4*. This protein is like an opsin protein in some ways since it contains within it another photon-absorbing molecule, called *FAD*. Unlike an opsin however, when the FAD inside Cry4 absorbs a photon, it becomes unstable and reacts with nearby chemical groups of the Cry4 protein, creating two new highly reactive molecules called *free radicals*. The chemistry is complex here, but the important part is that free radicals can be influenced by magnetic forces. Nobody knows yet what happens next, but some good guesses would be that the free radicals start a cell signaling reaction, much like the opsin initiates GPCR signaling in the eye, or that the Cry4 protein somehow directly opens an ion channel. As I said, nobody knows yet but most likely at some point in the future we will unravel the mystery.

A bold claim was made recently that a magnetic receptor protein complex had been found and is present in animals from fruit flies to mammals and even in humans [125]. This receptor complex has several features that definitively sound like the sort of thing we would expect of a magnetic field sensor protein. First, it contains iron and is able to orient or position itself within Earth's magnetic field. Second, this receptor particle is constructed of multiple iron-containing modules that are attached end-to-end, much like bacterial magnetosomes. Third, it also contains the Cry4 protein that I mentioned above. If this is all correct, this particle, called *MagR*, could form the basis for magnetoreception much like opsins and retinal form the basis of photon detection in the eye. How it might physically detect magnetic fields and pass this information on to a cell, converting magnetism into an electrical signal, is not known.

So, now you have a quick understanding of the predominant hypotheses for how a protein might be able to sense the Earth's magnetic field. These hypotheses have been argued back and forth in scientific circles for decades now, with no clear winner. But that's okay, especially since the first described magnetic field receptor combines both the magnetic particle hypothesis and the Cry4/free radical hypothesis into one tidy package so there's no need to argue about who's right! Maybe both hypotheses are correct! Or, you never know, maybe both are wrong?

Not everything is resolved yet because other scientists have not yet been able to confirm that MagR is a magnetic field receptor. The first attempt at this was to transplant the MagR complex into cells and see if it makes these cells depolarize in response to magnetic fields. It didn't [126]. Confirmation of findings by a second laboratory is a critical and necessary step in determining if any result is truly accurate, but this hasn't happened for MagR yet. The scientists who proposed MagR as a magnetic field receptor undoubtably did so believing their data was correct, and it might be correct. There could be many reasons why other labs have not been able to confirm MagR to be a magnetic field receptor complex. For instance, it's very possible that there is a channel protein that the MagR complex connects to that causes depolarization that is not present in other tissues. Another possibility is that in biology there are a tremendous number of unknowns. What this means is that even though scientists try their best to perform experiments that are perfectly controlled and the exact same every time, this is really difficult to achieve when you are working with living things that can change in unknown ways on a daily or even hourly basis. Even cells cultured in incubators wherein a lot of biological research happens, are extremely complex and it is impossible to make 100% sure that experimental conditions are 100% identical from day to day. This difficulty is likely a big part of the "Reproducibility Crisis" that is a worry in biological science. This crisis is based on the fact that it is unfortunately not uncommon for scientists to have difficulty reproducing other scientists results in their own labs. It's a major challenge for science that will need to be solved. Fortunately, there are some experimental approaches that are quite definitive. One very powerful technique scientists can use to prove that something works how they think it does is to delete that thing from an organism and see what happens. In the case of MagR, instead of trying to introduce MagR into cells that might be lacking an accessory protein needed for cellular depolarization, the critical experiment that needs to be done is to delete MagR from a magnetic field sensing animal and determine if the animal can still sense magnetic fields. At the time I am writing this book, this experiment has yet to be reported so I can't tell you the result. Undoubtably however, in the next year or so, we will be able to read all about it and hopefully, finally say that we understand the molecular biology of the final sensory frontier.

Chapter summary

In this chapter you have learned about what many people feel are the "final frontiers" of sensory biology. In the next few years, scientists will probably come to better understand how TRPA1 senses infrared radiation and will probably discover or confirm the identity of the magnetic field receptor. One of the most important points about this chapter is the idea that in some cases animals have mechanisms to sense things that we cannot. Things like infrasonic sounds, ultrasonic sounds, infrared and ultraviolet radiation, electric fields, and some chemicals in the air and water can go undetected by humans but can cause real problems for some animals that we share the planet with. Hopefully, as we learn more about these animal sensitivities, we will find ways to modify human development and behavior so we can better co-exist with our animal friends.

Chapter 10: Hidden Lessons

You did it! You finished the book. If you started this journey knowing virtually nothing about molecular biology, you hopefully now have a basic understanding of molecular function and an appreciation for how molecules control your life. You have a functional understanding that you can use to understand how your world works and, hopefully, appreciate more fully the living world around you. Maybe after finishing the book you'll be able to make good SWAG about the molecular world and impress your friends?

The sad truth though is that years from now you might not remember many of the specifics about what you learned here. You will probably not remember the details about how hair cells in your ear work. Maybe you will forget what Piezo proteins do... maybe you already have? Perchance you will only have a vague recollection about how GPCRs work. The good news is that this is all perfectly fine! All these fun examples were only tools to help you learn the basic principles of molecular biology. This is how most knowledge in science works. You remember the big picture and some of the details, but most of the details go into that dark place where lost memories go. It's perfectly normal and it happens to even the most intelligent people I know. What will not leave you, however, is the bigger understanding of how molecular biology works. These big ideas are the mile wide and inch deep river of knowledge that you keep for a lifetime.

The paradox of biology is that, of the natural sciences (geology, chemistry, physics, astrology, and biology), it is simultaneously the least and most complex. What I have presented in this book are some of the basic principles of biology. At their core, these principles are some of the simplest scientific concepts. They are comfortable, or even easy, to tackle. In contrast, the founding principles of chemistry and physics are much more challenging to get a general grasp of, and this is why I say that biology is (or can be) the simplest natural science. But, as I also said, biology can be the most challenging. The reason for this, in a single word is, evolution. Chemistry and physics are not living systems in the same way that biology is. Evolution does not change the way atoms are built or the way chemicals interact. Evolution has no power over the fundamental physical forces of the universe. In fact these fundamental forces define every aspect of biological systems. Biological molecules must follow the principles of physics and chemistry. But evolution of biological systems produces an endless variety of molecules. Nobody can understand this great complexity of biology in its entirety. As a friend of mine says "biology has too many undefined moving parts", and I tend to agree with him on this.

The good news, however, is that you don't need to remember all the moving parts in all their gory detail. You only need to understand the general idea of how they work individually and how they work together. For most of the biologists I know, the best we can do is develop our river of knowledge but also have a few very deep spots in our rivers. These deep spots represent our specialties. We learn as much as we can about an incredibly specialized corner of biology and become the expert in that little world. We have a broad understanding of the rest of biology but like anybody else, when we need to know the details of something that is not in our specialty, we read.

I have given you lots of examples of how the molecular world works. I tried to use fun examples that most people have some experience with and can relate to. Hopefully you enjoyed learning how drugs, toxins, eels, and your senses work. Hidden within those examples were the principles, the fundamentals, or the big ideas (call them what you want) that I think form the core of what I really wanted to get at with this book. In this last section, I'm going to directly discuss the big ideas that were hidden in the text and remind you about where you encountered that big idea. I'll expand on these ideas a little bit and hopefully help you remember them for a long time.

Principle 1: Molecular shape and chemistry dictate interactions between any two molecules. This is one of the most essential of the fundamental lessons I hope you remember in the years to come. There are so many analogies that try to describe this principle. "Like a key in a lock", "like a puzzle piece in the puzzle", or sometimes "like Legos clicking together". No matter how you chose to remember this lesson, just remember that biology is driven by interactions between molecules. Function and form are inextricably linked. Some of the specific examples I used to make this point were: penicillin interacting with its receptor; the different structures of glucose, sucrose, and sucralose affecting receptor binding and sweetness; retinal changing shape and inducing the subsequent shape changes in the corresponding opsin and G-protein; and capsaicin changing the shape of TRPV1. These are just a few examples to remind you of the fact that ligands bind to receptors if, and only if, the shapes and chemistries are a match. When they do bind, the ligands often change the function of the receptor. The next time you take a medicine, think about how the ligand in the pill you are swallowing is changing the shape and function of some receptor somewhere in your body.

Principle 2: In biology, there is always an exception. To be perfectly accurate, there are a few things in biology that (to the best of my knowledge) are constant and do not have exceptions. First, all cells on Earth have a cell membrane. Second, chaos (entropy or disorder) is always increasing. Third, biology is always evolving. Try as I might, these are the only constants I can think of in biology. Most everything else in biology has exceptions. Rules that seem like they are set in stone and invariable will sometimes be bent or broken. Here I taught you that the active sites of most receptors are built such that they accept a single ligand. This is important to maintain specificity between molecular interactions. And yet there are exceptions to this rule. For example, smell and taste receptors seem to have active sites that can bind to a variety of ligands. Some medicines can have side effects by interacting with receptors that they are not supposed to. And TRPA1 has a wide range of ligand shapes. So, even when you think you understand something really well, biology can still surprise you with something new. It's part of what keeps the science so interesting and frustrating at the same time.

Why so many exceptions? Because evolution is, as Richard Dawkins might say, "a blind watchmaker". It is restrained only in that its creations, if they are to be successful, must survive. A broken watch has no use, but as long as a watch can still tell time, even if some of its pieces are broken, it is useful and won't be discarded. A broken sensory system has no use, but as long as it can tell your brain something about its environment, it is useful. Because of this, endless variations have survived because they at least sort of work, sometimes in unanticipated ways.

Evolution does not strive for "perfection". It only makes

functional solutions to common problems. Think about humans. Are we perfect beings? I would say not because we are still prone to a host of diseases, malformations, deficiencies, etc. We are not perfect, but our physiology is "good enough" to allow us to survive long enough to make babies. Personally, I like to think of evolution in terms of the U.S. tax code. It originally started out as something very simple and straightforward. But here and there, little changes were made. A tax haven over here, a loophole over there. Maybe little changes that help the tax code change and adapt with the times. After passing through the many hands of many politicians, each with their own special interests, mind you, it barely resembles what it started out as in the first place and it can be really confusing. It evolved. Despite all this, the most important thing, the only thing that matters, is that it still works! Does it mean that it is optimized and perfect for what it does? Definitively not. But the strengths and weaknesses of the system (hopefully) balance each other out and in the end, we have a workable solution. Evolution does not always arrive at the "perfect" creation either, but rather a creation that simply works. The blind watchmaker doesn't always make a perfect watch. His watches just tell time and that's enough to keep them out of the trash can for a little while longer. If you can hold on to this understanding of evolution, you will be able to understand why there are so many exceptions in biology.

Principle 3: Ions cannot pass through membranes alone; they need a channel. One of the few rules in biology that has not yet been broken (as far as I know), is that all cells on Earth require a cell membrane. The most important function of the cell membrane is to define and separate the inside and outside of the cell. Despite this, ions like sodium, potassium, and calcium need to move in and out of cells for cells to live and to carry out their functions. Since cell membranes are very good at blocking ions from moving through them, the only way to make this happen is to have specific channel proteins that open and close in response to specific ligands or environmental stimulations. Specific examples that we discussed include channel proteins like TRPV1, Peizo, Na_v1.7, Ca_v1.3, and TRPA1. This is the fundamental basis on which all sensory systems are based. Moreover, many of the drugs and toxins that humans can be exposed to function by either increasing or decreasing channel function. If you can hold on to a good understanding of basic channel function, you will be able to understand at least a little bit of how all sensory systems work.

Principle 4: Cells have internal dialogs through their cell signaling systems. Cell signaling is one of the most complex and confusing topics in all of biology. The example I provided was the G-protein-coupled receptor system that plays important roles in vision, smelling, and tasting. G-protein-coupled receptors are important for a wide variety of functions, are the most common receptor proteins in humans, and many of our pharmaceuticals work by binding to and changing the function of these receptors. There are many, many other signaling pathways that cells can use and often times, the communication pathways overlap, or engage in something called "cross-talk". Imagine a mountain of spaghetti noodles that are haphazardly piled up on a plate where each noodle represents a different communication pathway. The noodles are all intertwined and if you try to unravel the noodles and you will get a big knot of slippery noodles. Cell signaling is like that plate of spaghetti in many ways, but, unbelievably, somehow it all works! It is the fundamental way that cells communicate within themselves.

Many of the diseases that humans suffer from are caused by a breakdown in these communication pathways. In particular, one of the primary causes of cancer are the development of what are called "oncogenes", or cancer genes. To understand how oncogenes work, imagine what would happen if a GPCR were to get mutated in such a way that it always assumed its activated state. The receptor would be perpetually telling the cell that its ligand was bound to it even though the ligand might be nowhere near the GPCR. This is what oncogenes do. They "short circuit" cell communication pathways and fool the cell into thinking that it is always stimulated so they always grow. They even ignore the "stop signs" that usually make the cells slow down and stop growing. Because of the importance of cell signaling in human disease, many of the drugs we take are designed to modify these communication pathways. Keep the idea of cell signaling in mind the next time you take a medication since there is a good chance the molecules in that medicine will be tweaking a cell signaling pathway in your body.

Principle 5: Cells need receptors to respond to external stimuli. We take it for granted, but only some of our cells are capable of interacting with the external world. Why don't all your trillions of cells respond to light, or capsaicin, or sound, or whatever? Because not all your cells have all the receptors for all the stimuli in the world. They don't need to. It would be a waste of energy for your intestines to produce light receptors. Only some cells have the ability to sense the external world because only these cells have the right receptors. And it's a good thing too. Imagine if all your cells made all the possible receptors for all the possible stimulations! You would be tasting with your hands, smelling with your feet, and whatever other uncomfortable sensations you can imagination. No thanks, I think I'll keep my receptors right where they are supposed to be.

Principle 6: Biology has evolved many ways to detect the physical world. The world around us is filled with stimulations. Photons flying around, sound waves coursing through the air, smell molecules floating off of flowers and bread and corpses, sharp coffee table corners and hot coffee, and the list goes on and on. Creatures need to detect and respond to these things to navigate through the world and survive.

The only types of energy that I can think of that animals as a group cannot detect are the very short and very long wavelengths at the extreme ends of the electromagnetic spectrum. As far as I know, we have not found any animals that can detect short wavelength photons like gamma rays and X-rays, or long wavelength photons like radio waves and microwaves. For these, humans have invented machines and tools that extend our senses by converting those types of energy into sound waves or visible light; types of energy that we can detect.

But biology has evolved ways to sense and respond to pretty much every other kind of physical stimulation out there. Even things like magnetic fields, infrared radiation, electric fields, and ultrasonic sounds that humans cannot detect can be detected by various creatures like birds and snakes.

A big question in my mind is whether or not there are other stimulations or energies in the world that animals have not yet evolved to detect. Maybe there are hidden forms of energy just waiting to be discovered. Maybe this is the great unknown that lurks behind the velvet curtain of death? But then again, if these hidden energies really do exist, I would think that some biological creature would have already evolved to detect them.

As far as I am aware, this is not the case.

Principle 7: Cell biology and molecular biology work together. It is very common in biology to find examples of specialized cellular structures that support, enable, and cooperate with specific molecular biology functions. In many cases, this cooperation between cell structure and molecular function is essential for optimal performance. For example, the stereocilia and tip links of hair cells are required for the most efficient activation of the mechanoreceptors that enable hearing. Neurons with their extended axons, ligand activated channels, and voltage gated channels enable the release of synaptic vesicles at the far end of the neuron. Electrocytes with their strategically placed sodium and potassium channels, stacked up like batteries, enable the strong electric discharges of electric eels. Finally, rods and cones with their many layers of flattened disks to maximize photon absorption by opsins. Examples like these are endless.

Principle 8: Sensory systems need to be amplified. The amount of energy that most sensory systems detect is very small. The eardrum only needs to be deflected by less than a picometer to transmit sound into the ear. Rods can be activated by a single photon of light. Piezo channels only need to be distorted by a few nanometers to report pressure. To achieve this sensitivity, cells need to amplify the signals they receive. Just like a radio needs an amplifier to convert minute radio signals into sounds we can hear, our cells need amplifiers too. We discussed several of these amplifiers including neuronal depolarization, GPCR signaling, outer hair cells, the chain reaction of the clotting cascade and others.

Principle 9: Toxins are medicines, medicines are toxins. The famous Renaissance physician and father of toxicology, Paracelsus, is quoted as saying that "All things are poisons, for there is nothing without poisonous qualities. It is only the dose which makes a thing poison." More modern quotes have expanded this definition by clarifying that another difference between a poison and a medicine is sometimes the intended use! Regardless, I wager that Paracelsus would have been fascinated and amazed by our molecular understanding of the differences and similarities between toxins and medicines. Remember that all toxins and medicines are just ligands that bind to a receptor and change its activity. It doesn't matter if it comes from nature's kitchen or from a laboratory. Some people think that "natural" medicines are better than manufactured medicines even if they contain the exact same molecules. From the perspective of a biologist or a biochemist, this just doesn't make sense since, in the end, they are all just ligands.

Principle 10: There is no applied science without basic science. There are lots of examples of how government dollars are wasted on this project or that project. How "Professor Science" received a million dollar grant to study an insignificant and silly research question like how long worms live. Who cares about how long worms live, right? Some of these stories of wasted spending are definitively true, and some people love to point at these examples to support their position that the government shouldn't be spending money on research. Some people go even further and consider wasteful spending to include any research spending that does not directly apply to improving the human condition on Earth. What these people fail to understand is the difference between applied research and basic research and how applied research cannot exist without basic research.

Applied research includes those research projects that are trying to make a direct improvement on how well humans live. In the biomedical research world, examples of applied research might be developing new drugs for cancer treatment, new antibiotics, or new approaches to treating heart disease. Research topics that usually involve human subjects and have a clear application to human health. Basic research, on the other hand, is the foundation on which applied research is built. For example, if somebody is developing a new cancer drug, they will absolutely need to have a solid understanding not just of how cancer works, but also of how a cell works and how molecules interact with each other. This understanding of cells and molecules comes from basic research and it is essential to first perform basic research so that the applied research can build on top of it. Basic research is the base of the pyramid and applied science is the apex of the pyramid.

Remember that research takes money. Lots of it. It's difficult to convince private foundations or businesses to support basic research since there is often no immediate product on which they can capitalize. This is why government funding in the form of research grants is so important. The combined budgets for the major governmental funding agencies (The National Institute of Health and the National Science Foundation) were a little less than fifty billion dollars in 2020. Without this investment, basic research would essentially stop.

Of course, when that kind of money gets spent anywhere, by governments or private companies, there will always be some amount of wasted spending. But for every example of "bad" research spending, I guarantee you that there are far, far, far more examples of NIH- or NSF-supported projects that added a few little "bricks" of basic information to the base of the pyramid. Taken individually, these bricks are not much more than just cubes of baked clay, but these bricks are exactly what we need to build major advances in science and medicine. In the end, that million-dollar grant to "Professor Science" to study how long worms live might seem really strange, but believe it or not, worms share about 70% of their genes with humans, and aging is something that happens in all organisms. What we learn in worms about how they age is actually having a huge impact on our understanding of how humans age. Someday, that information is probably going to be used to lengthen human life spans and that seems like a pretty good use of basic research funding to me.

Principle 11: Bioelectricity is electricity. You have probably been told for a long time that our nerves are "electric". And now, hopefully you know why. Always remember the definition of electricity: the movement of charged particles. Neurons and electric eels make bioelectricity in fundamentally the same way. They separate charges across a cell membrane, then use channel proteins to move these charges across the cell membrane. That little journey those charges take across the cell membrane **is** bioelectricity. What sets the eels apart is that in their electric organs, the electrocytes do not have equal ion flow on all sides like nerves do. Instead, unequal flow of Na^+ and K^+ ions on opposite faces of the electrocyte produces a charge separation across the entire electrocyte. The electrocyte becomes a battery for a couple of milliseconds. Thousands of these mini batteries stacked up makes for a big discharge. If you still don't think this is electricity, ask the electric eel's dinner how it feels about it.

Principle 12: There are many things we do not know. Somedays it seems like there just aren't any good mysteries in science anymore. It seems like all of the easy and best questions have already been answered. Sometimes I get a great idea, only to discover later that somebody else has already done that experiment. In some ways, it is more difficult to be a scientist now than it was ten, fifty, or a hundred years ago. On the other hand, we have technologies that our scientific grandparents and great-grandparents could only dream of! We also have the wealth of information that has been gathered since humans first started recording their scientific results right at our fingertips. The truth is that many of the **obvious** questions in science have been answered. This is the nature of research and discovery, and of human nature. In most cases, each discovery builds on the previous one, and topics become more nuanced with each discovery. Eventually, we reach a point where it's difficult to understand how it all got started. But then, every once in a while, something is discovered that makes all of us scientists scratch our heads and think "Why didn't I think of that!" Something that seems to have been staring all of us right in the face, but nobody actually noticed. Until they did. A great example of this is the recent "RNA renaissance" where researchers discovered several new kinds of RNA and the amazing new biology that is possible with these new RNA molecules. These new findings are breaking the traditional rules that tell us what RNAs are and what they can do. There are still mysteries that remain to be discovered. One of the mysteries I mentioned previously is the nature of the magnetic field receptor that allows birds and turtles to sense and navigate by the Earth's magnetic field. Another is the identity of the mechanoreceptors in hair cells. These are fairly small mysteries in the context of the whole universe, but they are big enough that a scientist could make a career out of studying them. Are there really big discoveries yet to be made? Of course! If you are a scientist, you just need to know where to look and remember to keep your eyes, ears, and mind open to something new. Likely something you are not expecting.

Principle 13: Flexibility is fundamental to science. Imagine if you were trying to assemble a one-million-piece puzzle but didn't have any clue what the picture on the final assembled puzzle was supposed to look like. It sounds like a nightmare, right? As you started to assemble a few pieces, you might start to see a fragment of the picture and have an "ah-ha" moment, thinking that you know what the final puzzle is going to be. But then, as you keep working, you have another "ah-ha" moment and realize that what you were seeing before was only a small piece of the bigger picture, so you revise your thinking and keep building. This goes on and on, continually changing as you learn more about what you are building.

If you get overwhelmed and call up your friends to work on the puzzle with you, each person might start to make conclusions about the final picture that don't agree with each other. Irina thinks the puzzle is going to be a giraffe, Trudy thinks it's going to be a lion, and Ruth thinks it's going to be a unicorn. As you keep building, Audrey comes in and sees that the individual parts are going to connect together and show a picture of Noah's Ark.

I am not exaggerating when I say that this is how science works except that instead of a one-million-piece puzzle, science is an infinite-piece puzzle. You never know for sure what the final picture is going to look like. This sort of thing happens all the time in science, and it's unavoidable. It's the only way. As scientists, it's important that we don't speculate too much about what we don't know, especially when it comes to things like human health. As non-scientists, it's important to trust scientists, but also to understand the nature of science and to acknowledge that scientists are humans too. Sometimes they speak too soon, get things wrong, or are even sometimes tempted by money and power. In the end however, the scientific method will figure things out and the truth will emerge.

Principle 14: Evolution (and convergent evolution) is real and is still happening. We have discussed several examples of evolution throughout this book. Some of the most fascinating examples of evolution are those examples of convergent evolution. Ever since Darwin published *On the Origin of Species* in 1859, the theory of evolution has been gaining traction in the world. But there are still some individuals that either do not believe evolution is real or are perhaps a bit skeptical that evolutionary forces have shaped (and are still shaping) our world. One of the great things about evolution is that you can observe evolution in action by simply observing the world around you. And if you can't believe yourself, who can you believe?

A recent example of "modern" evolution from 2020 and into 2021 was of course the SARS-COV-2 virus that causes Covid-19. In the beginning of the pandemic, this virus was able to jump from an animal host (maybe a bat or a pangolin) into a human host. That first jump to humans was enabled by the evolutionary relationship between humans and other animals. Even though pangolins and humans look different on the outside, human genes are still very similar to those of a pangolin. This means that SARS-COV-2 probably used the same proteins in pangolins and humans that allow it to infect and grow in both species.

Now that SARS-COV-2 has infected the human population, it is acquiring new mutations that enhance its ability to grow in people. These new mutations are the variants like the N501Y, and the N501Y + E484K variants that developed early in the pandemic. This nomenclature indicates that, for example, the 501st amino acid of the SARS-2 spike protein originally was an "N" but has been mutated to a "Y", hence N501Y. Most recently, the "delta variant" with mutations at T478K, P681R and L452R is the predominant strain in the world. These mutations make the virus more infectious and/or more dangerous. One of the principle ideas of evolution is the "survival of the fittest". If the ability of mutant viruses makes them more able to infect humans and produce even more new viruses, then these mutant viruses are more "fit" than the first viruses. These new variants have evolved to better suit their new host. What's more is that these variant viruses appear to be independently evolving the same mutations in multiple countries. Like prestin, eye development, and electric fish, these viral variations are another example of convergent evolution.

There you have it, the evolution of SARS-COV-2 in only a couple years. This is not the only example of evolution happening right before our eyes either. Other examples are the evolution of antibiotic resistance in bacterial cells like MRSA, and the resistance of cockroaches to many of our pesticides. What viruses, bacteria, and cockroaches have in common is that they all reproduce very, very fast. In your own digestive system, there have been more generations of bacterial cells produced than there have been humans on our planet. With each new generation of these little nasties, there is a new chance for evolutionary change. You might ask whether or not humans are still evolving? The answer is yes, although our generation time is so much longer than viruses, bacteria, and cockroaches that our evolutionary process is much slower and as individuals, we don't even notice it. And yet, we are evolving. One of my favorite examples of human evolution in the last ~10,000 years is the story of lactate persistence, or the ability of human adults to digest milk. Believe it or not, adult humans have not always been able to digest milk. But sometime around 10,000 years ago, it seems that genetic mutations began to appear in the population that made some adults able to continuously produce the enzyme *lactase* that is normally only made in infants [127]. These people evolved to drink milk throughout their whole life. Lactase allows people to digest the natural sugar *lactose* in milk and If you lack lactase but drink milk, you will probably get severe diarrhea. Just ask anybody who is lactose intolerant! Adults with the lactase enzyme were probably to use milk as a food source and this helped them survive better and pass on their genes. That is interesting, but even more interesting is the fact that the ability to consume milk happened twice in human history. First, in Europe through one mutation, but again in people of Middle Eastern descent through a different mutation [128]. Two different mutations that cause the same effect. One last example of convergent evolution, but in humans this time!

There you have them. The little jewels that were hidden in the chapters. I hope you will take these big lessons with you into your future and pull them out when you are trying to understand how your world works. I mentioned in the beginning that sometimes we need a "tour guide" to help navigate complex subjects. Somebody to hold your hand and point out things that are fascinating but might go under the radar if you don't know what you are actually looking at. It has been my honor to hold your hand and be your tour guide for that journey.

The Last Word

As a professor who makes a living out of talking in front of people, I get really tired of hearing my own voice! I absolutely love it when students ask questions to break up the monotony of my lectures! Whether you knew it or not, from the first day you picked up this book, you were my student.

So, do you have questions for the professor? I want to open up discussions and answer your questions, whether they are about topics in this book or anything else about biology that you might have become curious about. I'm also very much interested in taking requests that readers would like to learn about in subsequent volumes. If you want to reach out for whatever reason, you can reach me on my Facebook page @ questionsfortheprofessor. Stop by and say hi!

Finally, I have discovered that it is very difficult to get the attention of book publishing companies, or even literary agents, if you are a first-time author. So, I decided to self-publish this book. My "marketing team" is me, myself, and I, and marketing the book is a "grass-roots" operation. If you enjoyed this book, please consider leaving a review on the Bookbaby.com (the publisher) website or maybe on Amazon.com, or telling your friends about the book, or both. I'm not picky but either will make a world of difference to me.

Finally, if you enjoyed this book, you might also enjoy the next book I am working on. It will be another guide for non-scientists that describes how our various organs and body systems function at the cellular and molecular levels.

Acknowledgements

I'd like to give a nod to the many people who helped me in one way or another with this project. First and foremost, I need to give my sincere thanks to (soon to be) Dr. Elise Overgaard. Elise is an incredibly talented person who volunteered to edit this book. I'm confident that without the perspective of Elise, I would have stuck my foot in my mouth on several occasions. Hopefully, I'll be able to compensate her for her time with more than wine in the near future? I'd also like to extend my sincere thanks to all the people who volunteered to read and comment on the first draft. Thank you also to Dr. Henry Charlier, who you might know better by his alias, Dr. Picklestein. Your conversations and suggestions while we were not catching fish during the pandemic were more helpful than you probably realize. A special thank you to Dr. Robert Harrison for generously letting me use his beautiful electron micrographs of various structures in the inner ear. The link to his Auditory Science Lab is https://lab.research.sickkids.ca/harrison. Thank you also to Dr. Peter Dallos (Northwestern University) for discussing inner ear function with me and Dr. Jian Payandeh (Genentech) for helping with understanding how ProTx-II modifies $Na_v1.7$ structure and function. I'd also like to thank Conner Patricelli and Daniel Fologea for helpful discussions about biophysics. Finally, I'd like to thank the many, many people who let me bounce ideas off them, who discussed interesting topics with me, and who generally helped me to settle on a table of contents.

About the author and illustrator

Dr. Allan Albig is a professor of biology of Boise State University where he runs a research lab and teaches classes in cell and molecular biology. When he's not thinking about science, Allan is probably fishing or dancing salsa and jive with his wife.

Roxanna (Roxy) Albig is a graphic illustrator and designer. She dabbles in digital and mural work but focuses her attention to her detailed ink and alcohol traditional drawings. Currently Roxy is enrolled in Boise State University and is working toward her degree in illustration and business. She enjoys park and dance roller skating, participating in the local Boise music scene (both as an artist and a musician), and of course camping and fly fishing with her dad! More of Roxy's work can be found on her Instagram page @avacadooverlord.

It's no coincidence that both Allan and Roxy have the same last name. This book is a collaboration between father and daughter that hopefully takes advantage of their collective skills.

References Cited

Okay, let's be honest... references are not the most exciting things in the world. But, in the world of science and science writing, references are essential to support the things an author says. Without a proper reference, how can you really know that what someone is writing is fact or opinion? The reference provides the critical link back to the actual experimental discovery. In this book, I have referenced and discussed only the most important papers I needed to make my points. There are many other papers that I chose not to reference, simply to keep things less overwhelming. If your curiosity is going wild and you want to look up more information on any of the topics I have presented, the best place to do it is on the website called PubMed (https://pubmed.ncbi.nlm.nih.gov). PubMed is a database of nearly all the health-related papers ever published. It is maintained through U.S. tax dollars by the National Library of Medicine.

1.   Shang, J., et al., *Structural basis of receptor recognition by SARS-CoV-2*. Nature, 2020. **581**(7807): p. 221-224.

2.   Fleming, A., *On the Antibacterial Action of Cultures of a Penicillium, with Special Reference to their Use in the Isolation of B. influenzæ*. British journal of experimental pathology, 1929. **10**(3): p. 226-236.

3.   Bush, K., *The evolution of beta-lactamases*. Ciba Found Symp, 1997. **207**: p. 152-63; discussion 163-6.

4.   Baym, M., et al., *Spatiotemporal microbial evolution on antibiotic landscapes*. Science, 2016. **353**(6304): p. 1147-51.

5.   Beck, W.D., B. Berger-Bächi, and F.H. Kayser, *Additional DNA in methicillin-resistant Staphylococcus aureus and molecular cloning of mec-specific DNA.* J Bacteriol, 1986. **165**(2): p. 373-8.

6.   Lannoy, N. and C. Hermans, *The 'royal disease'--haemophilia A or B? A haematological mystery is finally solved.* Haemophilia, 2010. **16**(6): p. 843-7.

7.   Bakhle, Y.S., *Conversion of angiotensin I to angiotensin II by cell-free extracts of dog lung.* Nature, 1968. **220**(5170): p. 919-21.

8.   Koh, C.Y. and R.M. Kini, *From snake venom toxins to therapeutics--cardiovascular examples.* Toxicon, 2012. **59**(4): p. 497-506.

9.   Caterina, M.J., et al., *The capsaicin receptor: a heat-activated ion channel in the pain pathway.* Nature, 1997. **389**(6653): p. 816-24.

10. Cao, E., et al., *TRPV1 structures in distinct conformations reveal activation mechanisms.* Nature, 2013. **504**(7478): p. 113-8.

11. Bae, C., et al., *Structural insights into the mechanism of activation of the TRPV1 channel by a membrane-bound tarantula toxin.* Elife, 2016. **5**.

12. Yang, S., et al., *A pain-inducing centipede toxin targets the heat activation machinery of nociceptor TRPV1.* Nat Commun, 2015. **6**: p. 8297.

13. Geron, M., A. Hazan, and A. Priel, *Animal Toxins*

Providing Insights into TRPV1 Activation Mechanism. Toxins (Basel), 2017. **9**(10).

14. Andreev, Y.A., et al., *Analgesic compound from sea anemone Heteractis crispa is the first polypeptide inhibitor of vanilloid receptor 1 (TRPV1).* J Biol Chem, 2008. **283**(35): p. 23914-21.

15. Chu, Y., B.E. Cohen, and H.H. Chuang, *A single TRPV1 amino acid controls species sensitivity to capsaicin.* Sci Rep, 2020. **10**(1): p. 8038.

16. McKemy, D.D., W.M. Neuhausser, and D. Julius, *Identification of a cold receptor reveals a general role for TRP channels in thermosensation.* Nature, 2002. **416**(6876): p. 52-8.

17. Nilius, B. and A. Szallasi, *Transient Receptor Potential Channels as Drug Targets: From the Science of Basic Research to the Art of Medicine.* Pharmacological Reviews, 2014. **66**(3): p. 676-814.

18. Yang, Y., et al., *Mutations in SCN9A, encoding a sodium channel alpha subunit, in patients with primary erythermalgia.* J Med Genet, 2004. **41**(3): p. 171-4.

19. Cox, J.J., et al., *An SCN9A channelopathy causes congenital inability to experience pain.* Nature, 2006. **444**(7121): p. 894-8.

20. Schmidt, J.O., M.S. Blum, and W.L. Overal, *Hemolytic activities of stinging insect venoms.* Archives of Insect Biochemistry and Physiology, 1983. **1**(2): p. 155-160.

21.   Piek, T., et al., *Poneratoxin, a novel peptide neurotoxin from the venom of the ant, Paraponera clavata.* Comp Biochem Physiol C Comp Pharmacol Toxicol, 1991. **99**(3): p. 487-95.

22.   Johnson, S.R., et al., *A reexamination of poneratoxin from the venom of the bullet ant Paraponera clavata.* Peptides, 2017. **98**: p. 51-62.

23.   Yang, S., et al., *Discovery of a selective Na$_v$1.7 inhibitor from centipede venom with analgesic efficacy exceeding morphine in rodent pain models.* Proc Natl Acad Sci U S A, 2013. **110**(43): p. 17534-9.

24.   Xu, H., et al., *Structural Basis of Na$_v$1.7 Inhibition by a Gating-Modifier Spider Toxin.* Cell, 2019. **176**(4): p. 702-715.e14.

25.   Shen, H., et al., *Structures of human Na$_v$1.7 channel in complex with auxiliary subunits and animal toxins.* Science, 2019. **363**(6433): p. 1303-1308.

26.   Barchan, D., et al., *How the mongoose can fight the snake: the binding site of the mongoose acetylcholine receptor.* Proc Natl Acad Sci U S A, 1992. **89**(16): p. 7717-21.

27.   Drabeck, D.H., A.M. Dean, and S.A. Jansa, *Why the honey badger don't care: Convergent evolution of venom-targeted nicotinic acetylcholine receptors in mammals that survive venomous snake bites.* Toxicon, 2015. **99**: p. 68-72.

28.   Komives, C.F., et al., *Opossum peptide that can neutralize*

rattlesnake venom is expressed in Escherichia coli. Biotechnol Prog, 2017. **33**(1): p. 81-86.

29.   de Santana, C.D., et al., *Unexpected species diversity in electric eels with a description of the strongest living bioelectricity generator.* Nat Commun, 2019. **10**(1): p. 4000.

30.   Catania, K.C., *Electric Eels Concentrate Their Electric Field to Induce Involuntary Fatigue in Struggling Prey.* Curr Biol, 2015. **25**(22): p. 2889-98.

31.   Catania, K.C., *Electric eels use high-voltage to track fast-moving prey.* Nat Commun, 2015. **6**: p. 8638.

32.   Bellono, N.W., D.B. Leitch, and D. Julius, *Molecular basis of ancestral vertebrate electroreception.* Nature, 2017. **543**(7645): p. 391-396.

33.   Bellono, N.W., D.B. Leitch, and D. Julius, *Molecular tuning of electroreception in sharks and skates.* Nature, 2018. **558**(7708): p. 122-126.

34.   Gallant, J.R., et al., *Genomic basis for the convergent evolution of electric organs.* Science, 2014. **344**(6191): p. 1522-1525.

35.   Coste, B., et al., *Piezo1 and Piezo2 are essential components of distinct mechanically activated cation channels.* Science, 2010. **330**(6000): p. 55-60.

36.   Wang, L., et al., *Structure and mechanogating of the mammalian tactile channel PIEZO2.* Nature, 2019. **573**(7773): p. 225-229.

37.   Zhao, Q., et al., *The mechanosensitive Piezo1 channel: a three-bladed propeller-like structure and a lever-like mechanogating mechanism.* Febs j, 2019. **286**(13): p. 2461-2470.

38.   Poole, K., et al., *Tuning Piezo ion channels to detect molecular-scale movements relevant for fine touch.* Nat Commun, 2014. **5**: p. 3520.

39.   Woo, S.-H., et al., *Piezo2 is required for Merkel-cell mechanotransduction.* Nature, 2014. **509**(7502): p. 622-626.

40.   García-Mesa, Y., et al., *Merkel cells and Meissner's corpuscles in human digital skin display Piezo2 immunoreactivity.* J Anat, 2017. **231**(6): p. 978-989.

41.   Zeng, W.Z., et al., *PIEZOs mediate neuronal sensing of blood pressure and the baroreceptor reflex.* Science, 2018. **362**(6413): p. 464-467.

42.   Woo, S.H., et al., *Piezo2 is the principal mechanotransduction channel for proprioception.* Nat Neurosci, 2015. **18**(12): p. 1756-62.

43.   Marshall, K.L., et al., *PIEZO2 in sensory neurons and urothelial cells coordinates urination.* Nature, 2020.

44.   Kitazawa, T., et al., *Developmental genetic bases behind the independent origin of the tympanic membrane in mammals and diapsids.* Nature Communications, 2015. **6**(1): p. 6853.

45.   Qiu, X. and U. Müller, *Mechanically Gated Ion*

Channels in Mammalian Hair Cells. Frontiers in Cellular Neuroscience, 2018. **12**(100).

46.   Jia, Y., et al., *TMC1 and TMC2 Proteins Are Pore-Forming Subunits of Mechanosensitive Ion Channels.* Neuron, 2020. **105**(2): p. 310-321.e3.

47.   Kawashima, Y., et al., *Mechanotransduction in mouse inner ear hair cells requires transmembrane channel-like genes.* J Clin Invest, 2011. **121**(12): p. 4796-809.

48.   Indzhykulian, A.A., et al., *Molecular Remodeling of Tip Links Underlies Mechanosensory Regeneration in Auditory Hair Cells.* PLOS Biology, 2013. **11**(6): p. e1001583.

49.   Zheng, J., et al., *Prestin is the motor protein of cochlear outer hair cells.* Nature, 2000. **405**(6783): p. 149-155.

50.   Liberman, M.C., et al., *Prestin is required for electromotility of the outer hair cell and for the cochlear amplifier.* Nature, 2002. **419**(6904): p. 300-4.

51.   Menendez, L., et al., *Generation of inner ear hair cells by direct lineage conversion of primary somatic cells.* Elife, 2020. **9**.

52.   Davies, K.T.J., I. Maryanto, and S.J. Rossiter, *Evolutionary origins of ultrasonic hearing and laryngeal echolocation in bats inferred from morphological analyses of the inner ear.* Frontiers in Zoology, 2013. **10**(1): p. 2.

53.   Li, G., et al., *The hearing gene Prestin reunites echolocating bats.* Proc Natl Acad Sci U S A, 2008.

105(37): p. 13959-64.

54.   Liu, Y., et al., *Convergent sequence evolution between echolocating bats and dolphins.* Curr Biol, 2010. **20**(2): p. R53-4.

55.   Liu, Z., et al., *Parallel sites implicate functional convergence of the hearing gene prestin among echolocating mammals.* Mol Biol Evol, 2014. **31**(9): p. 2415-24.

56.   Corcoran, A.J., J.R. Barber, and W.E. Conner, *Tiger Moth Jams Bat Sonar.* Science, 2009. **325**(5938): p. 325-327.

57.   Neil, T.R., et al., *Moth wings are acoustic metamaterials.* Proc Natl Acad Sci U S A, 2020.

58.   Kimura, T., et al., *Guanine crystals regulated by chitin-based honeycomb frameworks for tunable structural colors of sapphirinid copepod, Sapphirina nigromaculata.* Scientific Reports, 2020. **10**(1): p. 2266.

59.   Land, M.F., *Image formation by a concave reflector in the eye of the scallop, Pecten maximus.* J Physiol, 1965. **179**(1): p. 138-53.

60.   Palmer, B.A., et al., *The image-forming mirror in the eye of the scallop.* Science, 2017. **358**(6367): p. 1172-1175.

61.   Ingram, N.T., A.P. Sampath, and G.L. Fain, *Why are rods more sensitive than cones?* J Physiol, 2016. **594**(19): p. 5415-26.

62.   Cronin, T.W. and M.J. Bok, *Photoreception and vision*

in the ultraviolet. The Journal of Experimental Biology, 2016. **219**(18): p. 2790-2801.

63.   BOETTNER, E.A. and J.R. WOLTER, *Transmission of the Ocular Media.* Investigative Ophthalmology & Visual Science, 1962. **1**(6): p. 776-783.

64.   Kojima, D., et al., *UV-sensitive photoreceptor protein OPN5 in humans and mice.* PLoS One, 2011. **6**(10): p. e26388.

65.   Gruener, A., *The effect of cataracts and cataract surgery on Claude Monet.* Br J Gen Pract, 2015. **65**(634): p. 254-5.

66.   Porter, M.L., et al., *The evolution of complexity in the visual systems of stomatopods: insights from transcriptomics.* Integr Comp Biol, 2013. **53**(1): p. 39-49.

67.   Futahashi, R., et al., *Extraordinary diversity of visual opsin genes in dragonflies.* Proceedings of the National Academy of Sciences, 2015. **112**(11): p. E1247-E1256.

68.   Colbourne, J.K., et al., *The Ecoresponsive Genome of Daphnia pulex.* Science, 2011. **331**(6017): p. 555-561.

69.   Thoen, H.H., et al., *A different form of color vision in mantis shrimp.* Science, 2014. **343**(6169): p. 411-3.

70.   Williams, D.L., *Light and the evolution of vision.* Eye (Lond), 2016. **30**(2): p. 173-8.

71.   Liu, X., S. Ramirez, and S. Tonegawa, *Inception of a false memory by optogenetic manipulation of a hippocampal*

memory engram. Philos Trans R Soc Lond B Biol Sci, 2014. **369**(1633): p. 20130142.

72.   Zhao, G.Q., et al., *The receptors for mammalian sweet and umami taste.* Cell, 2003. **115**(3): p. 255-66.

73.   Tu, Y.H., et al., *An evolutionarily conserved gene family encodes proton-selective ion channels.* Science, 2018. **359**(6379): p. 1047-1050.

74.   Maehashi, K., et al., *Bitter peptides activate hTAS2Rs, the human bitter receptors.* Biochem Biophys Res Commun, 2008. **365**(4): p. 851-5.

75.   Max, M., et al., *Tas1r3, encoding a new candidate taste receptor, is allelic to the sweet responsiveness locus Sac.* Nat Genet, 2001. **28**(1): p. 58-63.

76.   Montmayeur, J.P., et al., *A candidate taste receptor gene near a sweet taste locus.* Nat Neurosci, 2001. **4**(5): p. 492-8.

77.   Nelson, G., et al., *Mammalian sweet taste receptors.* Cell, 2001. **106**(3): p. 381-90.

78.   Yang, L., M. Cui, and B. Liu, *Current Progress in Understanding the Structure and Function of Sweet Taste Receptor.* Journal of Molecular Neuroscience, 2021. **71**(2): p. 234-244.

79.   Nie, Y., et al., *Distinct contributions of T1R2 and T1R3 taste receptor subunits to the detection of sweet stimuli.* Curr Biol, 2005. **15**(21): p. 1948-52.

80. Koizumi, A., et al., *Human sweet taste receptor mediates acid-induced sweetness of miraculin.* Proc Natl Acad Sci U S A, 2011. **108**(40): p. 16819-24.

81. Sun, H.J., et al., *Functional expression of the taste-modifying protein, miraculin, in transgenic lettuce.* FEBS Lett, 2006. **580**(2): p. 620-6.

82. Sun, H.J., et al., *Genetically stable expression of functional miraculin, a new type of alternative sweetener, in transgenic tomato plants.* Plant Biotechnol J, 2007. **5**(6): p. 768-77.

83. Kurihara, K. and L.M. Beidler, *Taste-modifying protein from miracle fruit.* Science, 1968. **161**(3847): p. 1241-3.

84. Fowler, A., *The miracle berry,* in *BBC News Magazine.* 2008.

85. Kearns, C.E., L.A. Schmidt, and S.A. Glantz, *Sugar Industry and Coronary Heart Disease Research: A Historical Analysis of Internal Industry Documents.* JAMA Internal Medicine, 2016. **176**(11): p. 1680-1685.

86. McGandy, R.B., D.M. Hegsted, and F.J. Stare, *Dietary fats, carbohydrates and atherosclerotic vascular disease.* N Engl J Med, 1967. **277**(5): p. 245-7 concl.

87. O'Connor, A., *How the Sugar Industry Shifted Blame to Fat,* in *The New York Times.* 2016.

88. KENNEDY, C.F.A.B., *Public confidence in scientists has remained stable for decades,* in *FACTTANK.* 2020.

89. Li, X., et al., *Human receptors for sweet and umami taste.* Proc Natl Acad Sci U S A, 2002. **99**(7): p. 4692-6.

90. Nelson, G., et al., *An amino-acid taste receptor.* Nature, 2002. **416**(6877): p. 199-202.

91. Zozulya, S., F. Echeverri, and T. Nguyen, *The human olfactory receptor repertoire.* Genome Biol, 2001. **2**(6): p. Research0018.

92. Bushdid, C., et al., *Humans Can Discriminate More than 1 Trillion Olfactory Stimuli.* Science, 2014. **343**(6177): p. 1370-1372.

93. Bilinska, K., et al., *Expression of the SARS-CoV-2 Entry Proteins, ACE2 and TMPRSS2, in Cells of the Olfactory Epithelium: Identification of Cell Types and Trends with Age.* ACS Chem Neurosci, 2020. **11**(11): p. 1555-1562.

94. Bryche, B., et al., *Massive transient damage of the olfactory epithelium associated with infection of sustentacular cells by SARS-CoV-2 in golden Syrian hamsters.* Brain Behav Immun, 2020. **89**: p. 579-586.

95. Hansen, A., *Olfactory and solitary chemosensory cells: two different chemosensory systems in the nasal cavity of the American alligator, Alligator mississippiensis.* BMC Neurosci, 2007. **8**: p. 64.

96. Buck, L. and R. Axel, *A novel multigene family may encode odorant receptors: a molecular basis for odor recognition.* Cell, 1991. **65**(1): p. 175-87.

97. Hussain, A., et al., *High-affinity olfactory receptor for the*

death-associated odor cadaverine. Proc Natl Acad Sci U S A, 2013. **110**(48): p. 19579-84.

98.   Shirasu, M., et al., *Chemical Identity of a Rotting Animal-Like Odor Emitted from the Inflorescence of the Titan Arum (Amorphophallus titanum).* Bioscience, Biotechnology, and Biochemistry, 2010. **74**(12): p. 2550-2554.

99.   Brodie, B., et al., *Bimodal cue complex signifies suitable oviposition sites to gravid females of the common green bottle fly.* Entomologia Experimentalis et Applicata, 2014. **153**(2): p. 114-127.

100.   Fischer-Tenhagen, C., et al., *A Proof of Concept: Are Detection Dogs a Useful Tool to Verify Potential Biomarkers for Lung Cancer?* Front Vet Sci, 2018. **5**: p. 52.

101.   Shirasu, M., et al., *Dimethyl Trisulfide as a Characteristic Odor Associated with Fungating Cancer Wounds.* Bioscience, Biotechnology, and Biochemistry, 2009. **73**(9): p. 2117-2120.

102.   Pu, D., et al., *Characterization of the key odorants contributing to retronasal olfaction during bread consumption.* Food Chem, 2020. **318**: p. 126520.

103.   Erbe, C., et al., *The Effects of Ship Noise on Marine Mammals—A Review.* Frontiers in Marine Science, 2019. **6**(606).

104.   Simonis, A.E., et al., *Co-occurrence of beaked whale strandings and naval sonar in the Mariana Islands,*

Western Pacific. Proc Biol Sci, 2020. **287**(1921): p. 20200070.

105.   Dominoni, D.M., et al., *Why conservation biology can benefit from sensory ecology.* Nat Ecol Evol, 2020. **4**(4): p. 502-511.

106.   Noble, G.K. and A. Schmidt, *The Structure and Function of the Facial and Labial Pits of Snakes.* Proceedings of the American Philosophical Society, 1937. **77**(3): p. 263-288.

107.   Gracheva, E.O., et al., *Molecular basis of infrared detection by snakes.* Nature, 2010. **464**(7291): p. 1006-11.

108.   Jordt, S.E., et al., *Mustard oils and cannabinoids excite sensory nerve fibres through the TRP channel ANKTM1.* Nature, 2004. **427**(6971): p. 260-5.

109.   Karashima, Y., et al., *TRPA1 acts as a cold sensor in vitro and in vivo.* Proc Natl Acad Sci U S A, 2009. **106**(4): p. 1273-8.

110.   Cordero-Morales, J.F., E.O. Gracheva, and D. Julius, *Cytoplasmic ankyrin repeats of transient receptor potential A1 (TRPA1) dictate sensitivity to thermal and chemical stimuli.* Proc Natl Acad Sci U S A, 2011. **108**(46): p. E1184-91.

111.   Corfas, R.A. and L.B. Vosshall, *The cation channel TRPA1 tunes mosquito thermotaxis to host temperatures.* eLife, 2015. **4**: p. e11750.

112. Lu, T., et al., *Odor coding in the maxillary palp of the malaria vector mosquito Anopheles gambiae.* Curr Biol, 2007. **17**(18): p. 1533-44.

113. DeGennaro, M., *The mysterious multi-modal repellency of DEET.* Fly (Austin), 2015. **9**(1): p. 45-51.

114. Bahia, P.K., et al., *The exceptionally high reactivity of Cys 621 is critical for electrophilic activation of the sensory nerve ion channel TRPA1.* J Gen Physiol, 2016. **147**(6): p. 451-65.

115. Mukhopadhyay, I., A. Kulkarni, and N. Khairatkar-Joshi, *Blocking TRPA1 in Respiratory Disorders: Does It Hold a Promise?* Pharmaceuticals (Basel), 2016. **9**(4).

116. Reese, R.M., et al., *Behavioral characterization of a CRISPR-generated TRPA1 knockout rat in models of pain, itch, and asthma.* Scientific Reports, 2020. **10**(1): p. 979.

117. Lin King, J.V., et al., *A Cell-Penetrating Scorpion Toxin Enables Mode-Specific Modulation of TRPA1 and Pain.* Cell, 2019. **178**(6): p. 1362-1374.e16.

118. Gui, J., et al., *A tarantula-venom peptide antagonizes the TRPA1 nociceptor ion channel by binding to the S1-S4 gating domain.* Curr Biol, 2014. **24**(5): p. 473-83.

119. Riera, C.E., et al., *Compounds from Sichuan and Melegueta peppers activate, covalently and non-covalently, TRPA1 and TRPV1 channels.* Br J Pharmacol, 2009. **157**(8): p. 1398-409.

120. Merkel, F.W. and W. Wiltschko, *Magnetismus und*

Richtungsfinden zugunruhiger Rotkehlchen (Erithacus rubecula). Vogelwarte, 1965. **23**(1): p. 71-77.

121. Wang, C.X., et al., *Transduction of the Geomagnetic Field as Evidenced from alpha-Band Activity in the Human Brain.* eNeuro, 2019. **6**(2).

122. Lohße, A., et al., *Functional Analysis of the Magnetosome Island in Magnetospirillum gryphiswaldense: The mamAB Operon Is Sufficient for Magnetite Biomineralization.* PLOS ONE, 2011. **6**(10): p. e25561.

123. Eder, S.H., et al., *Magnetic characterization of isolated candidate vertebrate magnetoreceptor cells.* Proc Natl Acad Sci U S A, 2012. **109**(30): p. 12022-7.

124. Diebel, C.E., et al., *Magnetite defines a vertebrate magnetoreceptor.* Nature, 2000. **406**(6793): p. 299-302.

125. Qin, S., et al., *A magnetic protein biocompass.* Nature Materials, 2016. **15**(2): p. 217-226.

126. Pang, K., et al., *MagR Alone Is Insufficient to Confer Cellular Calcium Responses to Magnetic Stimulation.* Frontiers in Neural Circuits, 2017. **11**(11).

127. Ségurel, L. and C. Bon (2017). "*On the Evolution of Lactase Persistence in Humans.*" Annual Review of Genomics and Human Genetics 18(1): 297-319.

128. Tishkoff, S. A., et al., (2007). "*Convergent adaptation of human lactase persistence in Africa and Europe.*" Nat Genet 39(1): 31-40.

Index